Spherical Means for PDEs

K.K. Sabelfeld and I.A. Shalimova

///VSP///
Utrecht, The Netherlands, 1997

VSP BV
P.O. Box 346
3700 AH Zeist
The Netherlands

First published in 1997

ISBN 90-6764-211-8

Printed in The Netherlands by Ridderprint bv, Ridderkerk.

Contents

List of Symbols

$\mathbb{R}^n$	Euclidean space of dimension n
G	a domain in $\mathbb{R}^n$
$\bar{G}$	the closure of G
$\Gamma = \partial G$	the boundary of G
$d(x)$	distance from a point x to the boundary Γ
$B(x,r)$	a ball of radius r centered at the point x
$S(x,r) = \partial B(x,r)$	a sphere of radius r centered at the point x
$K(x,r)$	a disk of radius r centered at x
$S(x,r) = \partial K(x,r)$	a circle of radius r centered at x
Ω	the unit sphere $S(0,1)$
$s_i = \frac{y_i - x_i}{R}$	coordinates of a unit directional vector s
$d\Omega$	the surface element of Ω
$dS_y = dS(y) = dS = do$	the surface element of $S(0,r)$ at a point y
$\Gamma(m)$	the Euler gamma-function
$\omega_m = 2\pi^{m/2}/\Gamma(m/2)$	the area of the surface of the unit sphere in $\mathbb{R}^m$
$N_r u(x) = N_r(u)$	the spherical mean of the function $u(x)$
$N_B\theta = B_x^r(\theta)$	the volume average of θ over a ball
$\binom{n}{k} = C_n^k$	binomial coefficients
$\Delta v_i = v_{i,jj}$	the Laplace operator of a function v_i
$\Delta^* = \mu\Delta + (\lambda + \mu)\,\mathrm{grad\,div}$	the Lamé operator
$\mathbf{u}$	the displacement vector, a solution to the Lamé equation $\Delta^*\mathbf{u} = 0$

$\varepsilon_{ij} = \frac{1}{2}(u_{i,j} + u_{j,i})$	the components of the strain tensor
$\tau_{ij} = 2\mu\varepsilon_{ij} + \lambda\delta_{ij}\,\mathrm{div}\,\mathbf{u}$	the components of the stress tensor
Π	the balayage operator
$J_\nu(z)$	the Bessel function of a complex z
$I_\nu(z) = i^{-\nu}J_\nu(iz)$	the modified Bessel function
$\mathrm{sh}\,z = \sinh(z)$	the hyperbolic sine function
$\mathrm{ch}\,z = \cosh(z)$	the hyperbolic cosine function
$Z(x,t)$	the fundamental solution of the heat equation
$n(y)$	a normal vector (exterior or interior) to the boundary at $y \in \Gamma$
λ, μ	the Lamé constants
$\alpha = (\lambda + \mu)/\mu$	
$\sigma = \frac{\lambda}{2(\lambda+\mu)}$	the Poisson constant
δ_{ij}	the Kronecker symbol
$\delta_x(y)$	uniform measure concentrated on $S(x,r)$
$\mathbb{E}\xi$	mathematical expectation of ξ.

Chapter 1

Introduction

It is well known that many boundary value problems for partial differential equations can be reformulated in the integral form. The classical example comes from potential theory where the original boundary value problems are reformulated through equivalent integral equations (on the boundary or in the volume, as in the Green formula, see, e.g., [4], [18], [20], [46]). The advantages of the integral formulation are well known: the solution is obtained on the whole, the boundary conditions are automatically taken into account, the questions of existence and uniqueness are resolved on the basis of the well developed Fredholm theory.

In this book, we deal with a new type of integral formulation of PDE's in the form of integral equations whose kernels are generalized functions. These equations are derived from the spherical mean value relations.

Let us recall the famous mean value relation for harmonic functions. Assume for simplicity that G is a bounded domain in $\mathbb{R}^n$, and let $S(x,r) \subset \bar{G}$ be a sphere of radius r centered at a point x. Let $u(x)$ be a regular harmonic function in G, i.e.,

$$\Delta u(x) = 0, \quad x \in G, \quad u \in C^2(G). \tag{1.1}$$

This means that the function $u(x)$ satisfies the Laplace equation in G and the second derivatives of u are continuous in G. Then for all $x \in G$ and all $S(x,r) \subset \bar{G}$ the following spherical mean value relation is true

$$u(x) = \frac{1}{\omega_n r^{n-1}} \int\limits_{S(x,r)} u(y) dS(y). \tag{1.2}$$

In the right-hand side here is the spherical mean of u, i.e., the integral of u over the sphere $S(x,r)$ with respect to the surface element measure $dS(y)$; ω_n is the surface area of a unit sphere in $\mathbb{R}^n$.

The following converse mean value relation is well known [4].

Theorem 1.1. Weak converse mean value theorem.

If a function $u(x) \in C^2(G)$ *satisfies the mean value relation* (1.2) *for all* $x \in G$ *and for all* $S(x,r) \subset \bar{G}$, *then* $u(x)$ *is harmonic in* G.

Less known is the following remarkable converse mean value theorem.

Theorem 1.2. The strong converse mean value theorem.

Assume that the Dirichlet problem

$$\Delta u(x) = 0, \; u|_\Gamma = \varphi, \; u \in C^2(G) \cap C(\bar{G}), \tag{1.3}$$

G *bounded, has a solution for any continuous function* φ. *Then if a function* $v \in C(\bar{G})$ *satisfies the mean value relation for at least one sphere* $S(x,r) \subset \bar{G}$, *for all* $x \in G$, *then* $v(x)$ *is harmonic in* G.

This exciting statement can be found in the book of Courant and Hilbert [4] where examples were given showing that all the conditions of the theorem are necessary. However, in [4], there was no attempt to derive from this approach an integral formulation of the Dirichlet problem.

In this book we deal with the integral formulations of the boundary value problems for various PDEs which we derive from the strong converse spherical mean value relations. In the second chapter we give the results for simple equations like the Laplace equation, the diffusion and Helmholtz equations, and the heat equation. The third chapter includes some high order equations (biharmonic, polyharmonic and metaharmonic equations), and Chapter 4 deals with some elliptic systems. Chapters 5, 6 and 7 are devoted to the Lamé equation, pseudo-vibration elastic and thermo-elastic equations.

It should be noted that different mean value relations can be derived: we can relate the spherical means of the solution and its derivatives to the values of these functions at the center of the sphere. There are many relations of this kind (e.g., see [50]) and we will use some of them. However the most interesting are the mean value relations which provide equivalent integral formulations of the original differential equation like the strong converse mean value theorem described above for the Laplace equation. Therefore our general objective is to give equivalent integral formulations of the corresponding PDEs.

In Chapter 8 we give some applications: these are probabilistic numerical algorithms for solving PDEs (the so-called Random Walk on Spheres algorithms) which we construct on the basis of the integral formulations. The solution in such methods is represented in the form of expectations over Markov processes generated in some sense by the spherical means. The algorithms are simple enough, and provide effective numerical solutions to the boundary value problems for complicated domains of high dimension. In addition, the implementation of the algorithms can easily be carried out for parallel computers.

The spherical mean value relations can be applied also to solving different problems of mathematical physics, for example, getting information about eigenvalues, constructing high-order finite difference schemes, finding asymptotics of the solutions, solving inverse problems of the potential theory, etc.

The probabilistic representations described in Chapter 8 assume minimum *a priori* information about the smoothness of the solution (e.g., the integral equation equivalent to the Laplace equation assumes only that the solution is continuous in $\bar{G}$). Actually an approach based on spherical mean value relations enables us to construct generalized solutions (see, e.g., [35]). Another interesting advantage of these representations is the possibility of solving numerically the exterior boundary value problems in complicated high-dimensional domains.

It should be noted that there is not much literature concerning spherical mean value relations for PDE's. First, there is the classical book of Courant and Hilbert [4]. In [50] (see also the references in this book) a short review of the works in this field has been carried out. The book [30] contains a set of mean value relations for metaharmonic equations. A series of articles by Diaz and Payne [5] deals with the mean value relations for elasticity problems. In [22], the mean value relations for some parabolic equations, and in [16], for telegraph equation were derived. We mention also the mean value relations where the integration is taken over other domains, for instance, ellipsoids [9].

There is also an interesting class of approximate mean value relations ([13], [35], [31]) which characterize the relevant equations. Special mean value relations are used in the Monte Carlo Random Walk on Spheres methods and can be found in the corresponding literature ([9], [40]).

Acknowledgement. This work was made possible in part by Grants from the Russian Foundation for Basic Research, N 94-01-00209 and N 96-01-01646, and INTAS-RFBR Grant N 95-IN-RU-726.

Chapter 2

Scalar second order PDEs

2.1. Spherical mean value relations for the Laplace equation and integral formulation of the Dirichlet problem

We consider equations in a domain G whose boundary ∂G is denoted by Γ and the closure by $\bar{G}$. Note that the domain G may be unbounded. Throughout the whole book we use the notations (partially already introduced in the previous chapter):

- $S(x,r)$ - a sphere of radius r centered at the point x,
- $B(x,r)$ - a ball of radius r centered at the point x,
- Ω - the unit sphere $S(0,1)$,
- $d\Omega$ - the surface element of Ω,
- $dS = do$ - the surface element of $S(0,r)$.
- ω_m - the area of the surface of the unit sphere in $\mathbb{R}^m$.
- $N_r u(x) = \frac{1}{\omega_m} \int\limits_{\Omega} u(x+rs)\, d\Omega(s)$ - the spherical mean of the function $u(x)$.

Let us start with a simple case, the Dirichlet problem for the Laplace equation in a domain $G \subset \mathbb{R}^m$:

$$\Delta u(x) = 0, \quad x \in G \ , \tag{2.1}$$

$$u(y) = \varphi(y), \quad y \in \Gamma \ . \tag{2.2}$$

We seek a regular solution to (2.1),(2.2), i.e., $u \in C^2(G) \bigcap C(\bar{G})$.

2.1.1. Direct spherical mean value relation

It is well known that every regular solution to (2.1) satisfies the spherical mean value relation:

$$u(x) = N_r u(x) := \frac{1}{\omega_m} \int\limits_{\Omega} u(x + rs)\, d\Omega(s) \tag{2.3}$$

for each $x \in G$ and for all spheres $S(x, r)$ contained in $\bar{G} := G \bigcup \Gamma$. The same is true for the volume mean value relation (it can be obtained directly from (2.3) by integrating):

$$u(x) = \frac{m}{\omega_m r^m} \int\limits_{B(x,r)} u(y)\, dy. \tag{2.4}$$

The mean value relation (2.3) can be derived by different methods. For small r, it is possible to use the method based on the power expansion of the integrand. We present this method here, and we will use it later to derive the mean value relations for different equations. The following statement is very useful, in particular, to get power expansions of the spherical means.

We denote by D the differential operator $D = \left(-i\frac{\partial}{\partial x_k}\right)_{1 \le k \le m}$ and $D^\alpha = D_1^{\alpha_1} \cdots D_m^{\alpha_m}$, where α is the multiindex:

$$\alpha = (\alpha_1, \ldots, \alpha_m), \alpha! = \alpha_1! \ldots \alpha_m!, D_k = -i\frac{\partial}{\partial x_k}, |\alpha| = \alpha_1 + \ldots + \alpha_m.$$

In [50], the following result is presented.

Lemma 2.1.

Let η be a measure in $\mathbb{R}^m$ with a compact support, and let $h(y) = \hat{\eta}$ be the Fourier transform of η:

$$h(y) = \hat{\eta} = \int\limits_{\mathbb{R}^m} \exp\{-i(y, s)\} d\eta(s).$$

Then

$$\int u(x+ry)\,d\eta(y) = \{h(-rD)\ u\}\ (x) \tag{2.5}$$

for x, r and u for which the left-hand side exists and the right-hand side converges.

Proof. The integral exists for sufficiently small $r > 0$ and as a function of r is analytic; the same for the right-hand side. Thus it is sufficient to prove the statement for sufficiently small r. The expansion of the integrand and the integration yield

$$\begin{aligned}
\int_{\mathbb{R}^m} e^{-i(y,s)}\,d\eta(s) &= \sum_{k=0}^{\infty} \frac{1}{k!} \int [-i(y,s)]^k\,d\eta(s) \\
&= \sum_{k=0}^{\infty} \frac{(-i)^k}{k!} \Big\{ \int \sum_{|\alpha|=k} \binom{k}{\alpha} y^\alpha s^\alpha\,d\eta(s) \Big\} \\
&= \sum_{k=0}^{\infty} \frac{(-1)^k}{k!} \Big\{ \sum_{|\alpha|=k} \binom{k}{\alpha} y^\alpha \int s^\alpha\,d\eta(s) \Big\} \\
&= \sum_{\alpha} \frac{1}{\alpha!} (-iy)^\alpha \Big\{ \int s^\alpha\,d\eta(s) \Big\}.
\end{aligned}$$

Here

$$\binom{k}{\alpha} = \frac{k!}{\alpha!(k-|\alpha|)!}\,.$$

Let

$$u(x+z) = \sum_\alpha a_\alpha(x) z^\alpha, \quad a_\alpha(x) = \frac{1}{\alpha!} \frac{\partial^{|\alpha|} u(x)}{\partial x_1^{\alpha_1} \cdots \partial x_m^{\alpha_m}}\,.$$

Then

$$\begin{aligned}
\int u(x+rs)\,d\eta(s) &= \int \Big\{ \sum_\alpha a_\alpha(x) r^{|\alpha|} s^\alpha \Big\} d\eta(s) \\
&= \sum_\alpha a_\alpha(x) r^{|\alpha|} \Big\{ \int s^\alpha d\eta(s) \Big\} \\
&= \sum_\alpha \frac{1}{\alpha!} (iD^\alpha)\, u(x)\, r^{|\alpha|} \Big\{ \int s^\alpha d\eta(s) \Big\}.
\end{aligned}$$

□

We consider the case when η is the uniform measure $d\Omega$. To formulate the result we define a function $W_m(\cdot)$ of the dimension m. Let $J_\nu(z)$ be the Bessel function:

$$J_\nu(z) = \sum_{k=0}^{\infty} \frac{(-1)^k}{\Gamma(k+\nu+1)\,\Gamma(k+1)} \left(\frac{z}{2}\right)^{2k+\nu}. \tag{2.6}$$

Here $\Gamma(\cdot)$ is the Euler Gamma-function. Let

$$W_m(z) = \Gamma\left(\frac{m}{2}\right)\left(\frac{2}{z}\right)^{m/2-1} J_{m/2-1}(z).$$

Theorem 2.1.

Let $u(x)$ be a real-valued analytic function. Then

$$N_r u(x) = \{W_m(ir\sqrt{\Delta})\}\; u(x) \tag{2.7}$$

for x, r and u for which the left-hand side is defined and the right-hand side exists.

Proof. We get (2.7) from (2.5) by choosing the measure η as $d\Omega/\omega_m$ and taking into account that the Fourier measure of $d\Omega/\omega_m$ is $W_m(|y|\,i)$, and $|D| = (-\Delta)^{1/2}$. □

For example, in $\mathbb{R}^3$

$$h(iy) = \frac{1}{4\pi}\int_0^{2\pi} d\varphi \int_0^{\pi} \sin(\theta)\, d\theta\; \exp(|y|\cos\theta) = \frac{\mathrm{sh}(|y|)}{|y|};$$

hence (2.5) has the form

$$N_r u(x) = \left\{\frac{\mathrm{sh}(r\sqrt{\Delta})}{r\sqrt{\Delta}}\right\} u(x). \tag{2.8}$$

Note that the last relation is the expansion in powers of the Laplace operator given in [4]

$$u(x) = N_r u(x) - \sum_{k=1}^{\infty} \frac{r^{2k}}{(2k+1)!}\, \Delta^k u(x) \tag{2.9}$$

since

$$W_m(ir\sqrt{\lambda}) = \sum_{k=0}^{\infty} \frac{\lambda^k\, r^{2k}}{2^k\; k!\; m(m+2)\dots(m+2k-2)}.$$

In $\mathbb{R}^2$ the "spherical" (a circular) mean has the representation

$$\frac{1}{2\pi}\int_0^{2\pi} u(x+re^{i\varphi})d\varphi = \{J_0(r\sqrt{-\Delta})\}\, u(x) = \sum_{k=0}^{\infty}(k!)^{-2}\Big(\frac{r}{2}\Big)^{2k}\;\Delta^k u(x).$$

Here $J_0(|x|)$ is the Bessel function which is obtained as the Fourier transform of the uniform measure $d\varphi/2\pi$.

Note that if we choose the uniform measure in the disk whose Fourier transform is $2J_1(|x|)/|x|$, we get

$$\frac{1}{\pi}\int_0^{2\pi}\int_0^1 u(x+rte^{i\alpha})\,tdt\,d\alpha = \sum_{k=0}^{\infty}\frac{1}{k!(k+1)!}\Big(\frac{r}{2}\Big)^{2k}\;\Delta^k u(x).$$

This relation can be generalized by choosing a general distribution ρ on the interval $[0,1]$ which leads to

$$\frac{1}{2\pi}\int_0^{2\pi}\int_0^1 u(x+rte^{i\alpha})\,d\rho(t)\,d\alpha = \sum_{k=0}^{\infty}(k!)^{-2}\Big(\frac{r}{2}\Big)^{2k}\;a_k\,\Delta^k u(x),$$

where

$$a_k = \int_0^1 (t/2)^{2k}\,d\rho(t).$$

Remark 2.1.

Note that the power expansions were given by Pizetti [32] (see also [4]): if a function $u(x) = u(x_1,\dots,x_m)$ belongs to $C^{2p+2}(G)$, $G \subset \mathbb{R}^m$, then for all $x \in G$ and all sufficiently small r

$$N_r u(x) = \sum_{k=0}^{p} r^{2k} C_k u(x) + Q_p(r)u(x),$$

and

$$N_B u(x) \equiv \frac{m}{\omega_m r^m} \int\limits_{B(x,r)} u(x+y)dy$$

$$= \sum_{k=0}^{p} r^{2k} C'_k u(x) + Q'_p(r)u(x),$$

where C_k and C'_k are the differential operators of the form

$$C_k = \frac{\Gamma(m/2)}{2^{2k} k! \Gamma(m/2+k)} \Delta^k ,$$

and

$$C'_k = \frac{m}{m+2k} C_k, \quad k \geq 0 .$$

The remainders can be estimated: for instance, in $3D$

$$Q_p(r)u(x) = \frac{1}{4\pi(2p+1)!} \int\limits_{B(x,r)} \frac{(r-|y|)^{2p+1}}{r|y|} \Delta^{p+1} u(x+y)dy ,$$

hence

$$|Q_p(r)u(x)| \leq c(p) r^{2p+2} \|u^{(2p+2)}\|_B ,$$

where the norm $\|u^{(2p+2)}\|_B$ is defined by

$$\|u^{(2p+2)}\|_B = \sum_{i_1,i_2,\ldots,i_\nu=1}^{m} \sup_{x \in B(x,r)} \left| \frac{\partial}{\partial x_{i_1}} \frac{\partial}{\partial x_{i_2}} \cdots \frac{\partial}{\partial x_{i_\nu}} u(x) \right| .$$

The spherical mean value relation for small radii for the harmonic functions follows from the obtained expansion (2.9). Generalization for all $S(x,r) \subset \bar{G}$ can be obtained by the analytical continuation, or by using the Green formula. Let us illustrate this in the case $\mathbb{R}^m$, $m \geq 3$. The Green formula reads

$$u(x) = \int\limits_{\partial G} \left\{ \mathcal{E}(x,y) \frac{\partial}{\partial \nu} u(y) - u(y) \frac{\partial}{\partial \nu} \mathcal{E}(x,y) \right\} do(y)$$

$$- \int\limits_{G} \Delta\, u(z)\, \mathcal{E}(x,z) dz.$$

Here

$$\mathcal{E}(x,y) = \frac{|x-y|^{2-m}}{(m-2)\omega_m} \,,$$

ν is the outward normal vector to the boundary ∂G, and

$$do(y) = r^{m-1} d\omega_m = \omega_m r^{m-1} d\Omega.$$

From this formula we immediately get the desired mean value relation, since $\partial/\partial\nu = \partial/\partial r$ for $G = B(x,r)$, $\Delta u = 0$ and

$$\int\limits_{S(x,r)} \frac{\partial}{\partial \nu} u \, do = 0.$$

The last equality follows from the well known Green formula

$$\int\limits_G (u\Delta v - v\Delta u)\, dx = \int\limits_{\partial G} \left(u\frac{\partial v}{\partial \nu} - v\frac{\partial u}{\partial \nu}\right) do \,,$$

which is true for arbitrary functions $u, v \in C^2(G)$.

We would now like to answer the questions:

Does the spherical mean value relation uniquely characterize the harmonic functions?

Is it possible to use the spherical mean value relation (in one form or another) to give an equivalent integral formulation of the Dirichlet problem for the Laplace equation?

How do we solve numerically the integral equation generated by the relevant spherical mean value relation?

In the next section we shall give answers to these questions.

In [59] the following general problem was studied: does the weak spherical mean value relation uniquely characterize the corresponding differential equation? We now present the relevant statements.

Assume that u is a real-valued analytic function which satisfies the mean value relation

$$u(x_0) = \int u(x_0 + ry) d\sigma(y) \tag{2.10}$$

where σ is a measure with compact support. Then, taking in Lemma 2.1 the measure $\eta = \sigma - \delta_{x_0}$, δ_{x_0} being the Dirac measure concentrated at the point x_0 and

$$h(z) = \sum_{\alpha} b_\alpha z^\alpha,$$

we get

$$\sum_{\alpha} (-1)^{|\alpha|} r^{|\alpha|} b_\alpha D^\alpha u(x) = 0. \tag{2.11}$$

From this, for sufficiently small $r > 0$, we get that

$$Q_j(D)u = 0, \quad j = 0, 1, 2, \ldots, \tag{2.12}$$

where

$$Q_j(D) = \sum_{|\alpha|=j} b_\alpha D^\alpha.$$

Thus the spherical mean value relation (2.10) for a real-valued analytic function holds if and only if the function u satisfies the infinite system of differential equations (2.12).

More generally, for continuous functions, the following statement holds (see [59]).

Theorem 2.2.

A function $u \in C(G)$, G bounded, satisfies the mean value relation (2.10) *for all $x \in G$ and all $0 < r < d(x)$ if and only if the function u is a weak solution to the system* (2.12).

It is interesting to find out when the system (2.12) is equivalent to a single equation $L(D)u = 0$. In this case σ is a distribution characterizing the operator $L(D)$.

Theorem 2.3.

The system (2.12) *is equivalent to a single equation of the type*

$$L(D)u = 0,$$

where $L(\xi)$ is a homogeneous polynomial, if and only if the polynomial $L(\xi)$ is divisible through all Q_j and there is an integer k such that $L = cQ_k$, c being a constant.

Finally, the form of the characterizing distribution is given in the following statement.

Theorem 2.4.

For each homogeneous polynomial L there exists a mass with a compact support such that a function $u \in C(G)$ is a weak solution to $L(D)u = 0$ if and only if the spherical mean value relation (2.10) *holds for all $x \in G$ and all $0 < r < d(x)$. Each mass of the form $L(D)m$ (m is a distribution with a compact support and $\hat{m}(0) \neq 0$) has this property.*

In the next section we deal with the strong converse mean value theorem which uniquely characterizes not only the equation but also the solution of the Dirichlet problem and provides an equivalent integral equation.

2.1.2. Converse mean value theorem

The mean value relation (2.3) characterizes the solutions to (2.1). We start with the weak converse mean value theorem, whose proof is well known (e.g., see [4]).

Proposition 2.1.

Suppose that a function $u(\cdot)$ is continuous in G and, for every sphere $S(x, r)$ contained in G, $u(\cdot)$ satisfies the mean value relation (2.3). *Then $u(\cdot)$ is harmonic in G.*

The proof follows from the expansion (2.9).

We formulate a stronger result which presents an equivalent formulation of the problem (2.1),(2.2).

Theorem 2.5. Strong Converse Mean Value Theorem.

Let $\varphi(\cdot)$ be a given continuous and bounded function on Γ; if G is unbounded, we suppose

(H) *the solution $u(\cdot)$ to the Dirichlet problem associated with $\varphi(\cdot)$ tends to 0 at infinity for the dimensions $m \geq 3$ and is bounded in the case $m = 2$.*

Assume that G is a domain for which the problem (2.1),(2.2) has a unique solution for any continuous and bounded function φ. Suppose that there exists a function $v \in C(G \bigcup \Gamma)$, $v|_\Gamma = \varphi$, such that the mean value relation (2.3) holds at every point $x \in G$ for at least one sphere $S(x,r) \subset G \bigcup \Gamma$; if G is unbounded, we suppose that $v(\cdot)$ tends to 0 at infinity in dimensions $m \geq 3$ and is bounded when $m = 2$.

Then $v(\cdot)$ is the unique regular solution to the problem (2.1),(2.2). The same conclusion holds if the volume mean value relation

$$v(x) = \frac{m}{\sigma_m r^m} \int\limits_{B(x,r)} v(y)\, dy$$

holds for every point $x \in G$ and at least one ball $B(x,r) \subset G \bigcup \Gamma$.

Proof. We use the same arguments as those given in [4], where the converse mean value theorem for the harmonic functions for bounded domains was given.

We shall now give the proof for the volume mean value relation (the same arguments work in the case of the spherical mean value relation).

Let u be the solution to the Dirichlet problem, and x be fixed in G. Since u satisfies the volume mean value relation for every ball $B(y,r)$ contained in G, we conclude that the function $u - v$ satisfies the volume mean value relation for $B(x,r)$. Let F be the set of points of G where $u - v$ attains its maximum M: even if G is unbounded, as $u - v$ tends to 0 at infinity (or is bounded, in dimensions $m = 2$), F is a compact set; let x_0 be a point of F whose distance to Γ is minimum. If x_0 were an interior point of G, we could find a ball $B(x_0, r_0) \subset G$ for which the volume mean value relation holds, and then $u - v$ would be equal to M inside $B(x_0, r_0)$, which is a contradiction with the definition of x_0. Therefore, x_0 belongs to Γ. We repeat the same argument for the minimum of $u - v$. Since, by hypothesis, $(u - v)|_\Gamma = 0$, we conclude $u \equiv v$ in G. □

Using this converse mean value relation we can formulate an integral equation equivalent to the Dirichlet problem.

2.1.3. Integral equation equivalent to the Dirichlet problem

In this section we use the **Strong Converse Mean Value Relation** to derive an equivalent integral equation of the second kind.

Let us denote by $d(x)$ the distance from a point $x \in G$ to the boundary Γ, and let

$$d^* = \sup_{x \in G} d(x) \ .$$

We introduce also an "ε-boundary" :

$$\Gamma_\varepsilon = \{x \in G : \ d(x) \leq \varepsilon\}.$$

Let

$$\delta_x(y) = \delta(|x-y| - d(x)).$$

For simplicity, we use this notation to indicate that $\delta_x(y)$ is equal to $1/\omega_m$ if $|x-y| = d(x)$ and to 0 otherwise. We define a kernel function k_ε by

$$k_\varepsilon(x,y) := \begin{cases} \delta_x(y) & \text{if } x \in G \setminus \Gamma_\varepsilon \ , \\ 0 & \text{if } x \in \Gamma_\varepsilon \ , \end{cases} \tag{2.13}$$

and define the integral operator K_ε by

$$K_\varepsilon \psi(x) := \int_G k_\varepsilon(x,y)\psi(y)dy \tag{2.14}$$

for each $\psi(\cdot) \in C(G)$.

We now fix the boundary function φ and suppose that the conditions of the converse mean value relation are satisfied. Denote by u the solution to the Dirichlet problem corresponding to φ, and by $f_\varepsilon(x)$ the function

$$f_\varepsilon(x) = \begin{cases} 0 & \text{if } x \in G \setminus \Gamma_\varepsilon \ , \\ u(x) & \text{if } x \in \Gamma_\varepsilon \ . \end{cases}$$

Consider the integral equation

$$v(x) = K_\varepsilon v(x) + f_\varepsilon(x) \ . \tag{2.15}$$

Note that the mean value relation implies (2.15). Thus if we assume that $u(x)$ is known in Γ_ε, (2.15) presents the desired integral equation of the second kind, and the converse mean value theorem states that this equation has a unique solution which coincides with the solution to the Dirichlet problem for the Laplace equation.

It remains to propose a method of calculation of the solution to (2.15). We show that the conventional successive iteration method applied to (2.15) is convergent.

Theorem 2.6.

For any $\varepsilon > 0$, the integral equation (2.15) has a unique solution given by the Neumann series

$$v(x) = f_\varepsilon(x) + \sum_{i=1}^{\infty} K_\varepsilon^i f_\varepsilon(x) \tag{2.16}$$

and it coincides with the solution to (2.1),(2.2) if the assumptions of the converse mean value relation are satisfied.

Proof. Let ε be fixed. It is simple to show the convergence of the series

$$f_\varepsilon(x) + \sum_{i=1}^{N} K_\varepsilon^i f_\varepsilon(x) \quad , \tag{2.17}$$

if $d^* < \infty$. It is then sufficient to prove the existence of $0 < \lambda < 1$ such that for any continuous and bounded function g

$$\|K_\varepsilon^2 g\|_{L_\infty} < \lambda \|g\|_{L_\infty}$$

(this fact also implies the uniqueness of the solution to (2.16)). Let

$$\nu(\varepsilon) = \varepsilon^2/(4d^{*2}) \quad . \tag{2.18}$$

For $x \in G \setminus \Gamma_\varepsilon$ we have

$$\int_G k_\varepsilon(x,y) \int_G k_\varepsilon(y,y')dy'\, dy = \int_{G\setminus\Gamma_\varepsilon} \delta_x(y) \left(\int_G \delta_y(y')dy' \right) dy$$

$$= \int_{G\setminus\Gamma_\varepsilon} \delta_x(y)dy \le 1 - \nu(\varepsilon) < 1.$$

Let $v(x) := f_\varepsilon(x) + \sum_{i=1}^{\infty} K_\varepsilon^i f_\varepsilon(x)$. It is clear that v satisfies

$$v(x) = K_\varepsilon v(x) + f_\varepsilon(x) \quad .$$

If x belongs to Γ_ε, then, for any y, $k_\varepsilon(x,y) = 0$ and thus $v(x) = f_\varepsilon(x) = u(x)$. On the other hand, if x belongs to $G \setminus \Gamma_\varepsilon$, then the definition of k_ε implies that v satisfies the mean value relation at x with the sphere of radius $d(x)$. We conclude by using the Theorem 2.5. □

Note that the condition $d^* < \infty$ is not necessary for the convergence of the iteration method. However it is difficult to study the convergence of the iteration method in this general case. In the last chapter we prove the convergence for the half-space $\mathbb{R}^3_+$ and the exterior of a sphere.

Remark 2.2.

In the integral equation (2.15) and in the iteration method we used the spheres $S(x, d(x))$. However the **Strong Converse Mean Value Relation** implies that in each step of the iteration method we can use a sphere of an arbitrary radius, say, r_x, such that $S(x, r_x) \subset \bar{G}$. In the last chapter we will see that this scheme with a special choice of r_x is reasonable when solving the Poisson equation by the Random Walk method.

2.1.4. Poisson–Jensen formula

More generally, a local integral equation can be derived if we replace the mean value relation by the Poisson formula representing the solution at an arbitrary point $x \in S(x_0, r) \subset \mathbb{R}^3$:

$$u(x) = \int_{S(x_0,r)} H(x,y)u(y)\, dS(y) \ , \tag{2.19}$$

where

$$H(x,y) = \frac{r^2 - x^2}{4\pi r|x-y|^3} \ ,$$

if we suppose that this equality holds for every $x_0 \in G$ at least for one sphere $S(x_0, r) \subset \bar{G}$, then $u(\cdot)$ is the unique regular solution to the Dirichlet problem.

In $2D$, there is a further generalization of the formula (2.19) – the so-called Poisson-Jensen formula [24].

Let $u(z) = u(r,\theta)$ be a univalent function harmonic in a disk $K(0,R)$ except for a set of singularities where u has logarithmic poles. We denote the non-zero poles by $\{\zeta_i\}$, $i = 1,\ldots,$ where the indexing is chosen so that $|\zeta_1| < |\zeta_2| < \ldots < |\zeta_i| < \ldots$. Let $\mu_j \ln|z-\zeta_j|$ be the principal part corresponding to ζ_j. It is convenient also to introduce the principal part $\mu_0 \ln|z|$ corresponding to the point $z=0$ taking $\mu_0 = 0$ if this point is not a logarithmic pole of $u(z)$.

Let us consider the conform mapping of the disk $|z| \le \rho$ on itself by

$$l_\zeta(z) = \rho^2 \frac{z-\zeta}{\rho^2 - \bar{\zeta} z},$$

where $|z| \le \rho < R$. This transformation maps the point ζ into the center of the disk. Since $l_\zeta(z)$ is an analytic function in the disk $|z| < \rho^2/|\zeta|$, having a single simple zero-point at $z = \zeta$ and $|l_\zeta(z)| = \rho$ at all points of the circle $|z| = \rho$, then the function

$$u_\zeta(z) = \ln\left[\frac{1}{\rho}|l_\zeta(z)|\right]$$

is harmonic if $|z| < \rho^2/|\zeta|$, $z \ne \zeta$ and has in this disk a logarithmic pole at the point $z = \zeta$ with the principal part $\ln|z-\zeta|$ and $u = 0$ on the boundary $z = \rho$. If one part of the logarithmic poles of $u(z)$, namely, the points $0, \zeta_1, \ldots, \zeta_{\nu(\rho)}$ lies inside this disk, and the rest is outside of it, then

$$v(z) = u(z) - \sum_{k=0}^{\nu(\rho)} \mu_k u_{\zeta_k}(z) = u(z) - \sum_{k=0}^{\nu(\rho)} \mu_k \ln\left|\frac{\rho(z-\zeta_k)}{\rho^2 - \bar{\zeta}_k z}\right|$$

is harmonic in all the points of a disk whose radius is larger than ρ (the radius is equal to the minimum of the two values: $|\zeta_{\nu(\rho)+1}|$ and $\rho^2/|\zeta_\nu(\rho)|$), while on the boundary $|z| = \rho$ this function coincides with $u(z)$. Therefore, for all $|z| = r < \rho$ we get

$$v(r,\theta) = \frac{1}{2\pi}\int_0^{2\pi} v(\rho,\alpha)\frac{r^2-\rho^2}{r^2+\rho^2-2r\rho\cos(\theta-\alpha)}\,d\alpha;$$

hence

$$u(r,\theta) = \frac{1}{2\pi}\int_0^{2\pi} v(\rho,\alpha)\frac{r^2-\rho^2}{r^2+\rho^2-2r\rho\cos(\theta-\alpha)}\,d\alpha + \sum_{k=0}^{\nu(\rho)} \mu_k \ln\left|\frac{\rho(z-\zeta_k)}{\rho^2-\bar{\zeta}_k z}\right| \tag{2.20}$$

which is known as the Poisson–Jensen formula. If the function has no logarithmic poles, then all the values μ_j are zeros, and we come to the Poisson formula.

2.2. The diffusion and Helmholtz equations

The aim of this section is to treat the diffusion equation

$$\Delta u(x) - \lambda u(x) = 0\ , \quad u|_\Gamma = \varphi\ , \tag{2.21}$$

where λ is a nonnegative constant, and the Helmholtz equation

$$\Delta u(x) + \lambda u(x) = 0\ , \quad u|_\Gamma = \varphi\ . \tag{2.22}$$

Using the function $W_m(z)$ introduced in the previous chapter we define a new function:

$$w_m(r,\lambda) = W_m(ir\sqrt{\lambda}).$$

Since $I_\nu(z) = J_\nu(iz)/i^\nu$, we get the expansion

$$\begin{aligned} w_m(r,\lambda) &= \Gamma\Big(\frac{m}{2}\Big)\Big(\frac{2}{r\sqrt{\lambda}}\Big)^{m/2-1} I_{m/2-1}(r\sqrt{\lambda}) \\ &= \Gamma\Big(\frac{m}{2}\Big)\sum_{k=0}^{\infty}\frac{\lambda^k r^{2k}}{2^{2k}\,k!\,\Gamma(k+m/2)} \\ &= \sum_{k=0}^{\infty}\frac{\lambda^k r^{2k}}{2^k\,k!\,m(m+2)\dots(m+2k-2)}\ , \end{aligned} \tag{2.23}$$

since

$$\Gamma\Big(k+\frac{m}{2}\Big) = \Gamma\Big(\frac{m}{2}\Big)m(m+2)\dots(m+2k-2)/2^k.$$

2.2.1. Diffusion equation

From (2.7) we deduce the mean value relation for the solution to (2.21):

$$u(x) = w_m^{-1}(r,\lambda)N_r u(x) \quad . \tag{2.24}$$

We obtain this mean value relation for arbitrary $S(x,r) \subset \bar{G}$ using a different approach.

Let

$$v(x,r) = N_r\, u = \frac{1}{\omega_m} \int\limits_{\Omega} u(x+sr)d\Omega(s).$$

Then the function $v(x,r)$ satisfies the Darboux equation [4]

$$\frac{\partial^2 v}{\partial r^2} + \frac{m-1}{r}\frac{\partial v}{\partial r} - \Delta\, v = 0$$

with the initial conditions

$$v(x,0) = u(x), \quad \frac{\partial v}{\partial r}(x,0) = 0.$$

Indeed, using the Gauss theorem we get

$$\begin{aligned} \frac{\partial v(x,r)}{\partial r} &= \frac{1}{\omega_m} \int\limits_{\Omega} \Big(\sum_{i=1}^m s_i \frac{\partial u}{\partial x_i}\Big)\, d\Omega \\ &= \frac{1}{\omega_m r^{m-1}} \int\limits_{S(x,r)} \frac{\partial u}{\partial \nu}\, do = \frac{1}{\omega_m r^{m-1}} \int\limits_{B(x,r)} \Delta u\, dy. \end{aligned} \tag{2.25}$$

Here

$$\frac{\partial}{\partial \nu} = \sum_{i=1}^m s_i \frac{\partial}{\partial x_i}$$

denotes the differentiation with respect to the outward normal vector to the sphere Ω, and dy is the volume element.

Differentiating once more yields the desired result:

$$\frac{\partial^2 v}{\partial r^2} = -\frac{m-1}{\omega_m r^m} \int\limits_{B(x,r)} \Delta u\, dy + \frac{1}{\omega_m r^{m-1}} \int\limits_{S(x,r)} \Delta u\, do = -\frac{m-1}{r}\frac{\partial v}{\partial r} + \Delta v.$$

The operator Δ and the spherical mean operator N_r are permutable [4]; therefore, from (2.21), we get $\Delta v = \lambda v$. Thus substituting $\Delta v = \lambda v$ in the Darboux equation we find that $v = N_r u(x)$ solves the problem

$$\frac{\partial^2 v}{\partial r^2} + \frac{m-1}{r}\frac{\partial v}{\partial r} - \lambda v = 0, \tag{2.26}$$

$$v(x,0) = u(x), \; \frac{\partial v}{\partial r}(x,0) = 0. \tag{2.27}$$

The solution to this problem is

$$v(x,r) = v(x,0) \cdot w_m(\lambda, r)$$

which is the desired spherical mean value relation.

In $G \subset \mathbb{R}^3$, we have from (2.8)

$$u(x) = \frac{\sqrt{\lambda} r}{\sinh(\sqrt{\lambda} r)} N_r u(x) \quad . \tag{2.28}$$

Note that (2.24) and (2.28) are true for complex λ such that $\mathrm{Re}\sqrt{\lambda} r < \pi$.

The diffusion equation is treated exactly according to the scheme described for the Laplace equation, since the integral formulation also can be given for (2.21).

Proposition 2.2. Integral Formulation.

We suppose that the Dirichlet problem for the diffusion equation has a unique solution for any continuous function φ; if G is unbounded, we suppose (H). Let $\varphi(\cdot)$ be a given continuous and bounded function on Γ. Assume that $u \in C(G \bigcup \Gamma)$ satisfies the mean value relation (2.24) *for each $x \in G$ at least for one sphere $S(x, r_x) \subset G$ and $v|_\Gamma = \varphi$. Then $u(\cdot)$ solves the Dirichlet problem* (2.21).

Proof. The proof repeats the arguments used in the case of the Laplace equation, since the strong maximum principle holds (this follows from the inequality $|w_m(r,\lambda)| \geq 1$ for nonnegative λ). □

2.2.2. Helmholtz equation

If $\lambda < 0$, we come to the Helmholtz equation.

We suppose in this case that $G \subset \mathbb{R}^3$ is a bounded domain and $0 > \lambda > \lambda_0(G)$ where $\lambda_0(G)$ is the principal eigenvalue of the Laplace operator in G.

Note that $\sqrt{\lambda} < \frac{\pi}{d^*}$; indeed, it is well known that, if $G_2 \subset G_1$, then $\lambda_0(G_2) \geq \lambda_0(G_1)$; in our case, for all $x \in G$, $S(x, d^*) \subset G$, and the eigenvalue $\lambda_0(S(x, d^*))$ is equal to $\frac{\pi^2}{d^{*2}}$.

Then the solution to (2.22) can be written in $\mathbb{R}^3$ in the form

$$u(x) = \frac{\sqrt{\lambda} r}{\sin(\sqrt{\lambda} r)} N_r u(x) \quad , \tag{2.29}$$

for any r such that $S(x,r)$ is included in G.

Under these restrictions the maximum principle is true (see [12]) and we conclude that the converse spherical mean value theorem is true.

Note that for the Helmholtz equation we can derive the volume mean value relation in a ball $B(x,r) \subset \mathbb{R}^n$

$$u(x) = \frac{1}{V_r \, 2^{n/2} \Gamma(\frac{n}{2}+1) \, \tau_{n/2}(\sqrt{\lambda} r)} \int\limits_{B(x,r)} u(y) dV(y) \ ,$$

where $V_r = \pi^{n/2} r^n / \Gamma(n/2+1)$ is the volume of the ball $B(x,r)$, and $\tau_\alpha(z) = z^{-\alpha} J_\alpha(z)$. We notice that the function $\tau_\alpha(z)$ is related to the function $W_\alpha(z)$ by

$$W_\alpha(z) = \Gamma(\alpha/2) 2^{\alpha/2-1} \tau_{\alpha/2-1}(z) \ .$$

The above volume mean value relation is obtained by integrating the spherical mean value relation for the Helholtz equation

$$u(x) = \frac{1}{W_n \omega_n} \int\limits_{\Omega} u(x+rs) \, d\Omega \ .$$

Remark 2.3.

The weak spherical mean value relations characterize the high-order elliptic equations with constant coefficients. Let us mention the following result (see, for example, [11] and [12]) which we briefly mentioned above.

Let μ be a nonnegative Borel measure with total mass equal to 1, such that the support of μ is contained in the unit sphere of $\mathbb{R}^m$ and not contained in any hyperplane. If u is a continuous function on some open set $G \subset \mathbb{R}^m$ having the mean value property

$$u(x) = \int u(x+ry)\mu(dy)$$

for every $x \in G$ and every positive $r < d(x)$, then $u \in C^\infty(G)$ and

$$\sum_{i_1+\ldots+i_m=n} A_{i_1\ldots i_m} \frac{\partial^n u}{\partial x_1^{i_1}\ldots \partial x_m^{i_m}} = 0, \quad n = 1, 2, \ldots,$$

where the coefficients are the moments

$$A_{i_1\ldots i_m} = \int x_1^{i_1}\ldots x_m^{i_m}\mu(dx).$$

Conversely, every infinitely differentiable solution of the last system of differential equations has the above mean value property.

2.3. Generalized second order elliptic equations

The expansions of the spherical mean of the type (2.9), (see also (3.10) below) suggest a device for defining a generalized Laplacian of nondifferentiable functions. For example, if in

$$\Delta v(x) = f$$

the functions v and f are merely continuous, one can say that $\Delta v = f$ in a generalized sense if

$$\lim_{r\to 0}\frac{1}{r^2}\Big[N_r v - v\Big] = \frac{f}{2m}\,.$$

An analogous point of view is reported in [35] and [48].

From an expansion which we will obtain in Chapter 5 it is possible to define generalized solutions to the Lamé equation

$$\mu\Delta \mathbf{u}(x) + (\lambda+\mu)\,\text{grad div}\,\mathbf{u}(x) = \mathbf{f}(x)$$

where $x \in \mathbb{R}^n$, $\mathbf{u} = (u_1, \ldots, u_n)$, as continuous functions satisfying the relation

$$\lim_{r\to 0}\frac{1}{r^2}\left[N_r^1\mathbf{u}-\mathbf{u}\right]=\frac{\mathbf{f}}{2(\lambda+\mu(n+1))}\ . \tag{2.30}$$

Here, the averaging matrix operator N_r^1 is defined by

$$\begin{aligned}\left(N_r^1\mathbf{u}\right)_i(x) &= \frac{1}{\omega_n}\int_\Omega (a+bs_i^2)u_i(x+rs)\ d\Omega \\ &+\frac{b}{\omega_n}\sum_{j\neq i}^{n}\int_\Omega s_is_ju_j(x+rs)\ d\Omega\ ,\quad i=1,\ldots,n\end{aligned}$$

where $a=1-\beta$, $b=n\beta$, and

$$\beta=\frac{(n+2)(\lambda+\mu)}{2(\lambda+\mu(n+1))}\ .$$

In [13], elliptic equations with nonconstant coefficients were considered:

$$Lu=0\ ,$$

where

$$L=\sum_{i,k=1}^{m}a_{ik}\frac{\partial^2}{\partial x_i\partial x_k}\ ,$$

and the coefficients are supposed merely measurable and bounded and the matrix $a(x)=(a_{ik}(x))$ symmetric and positive definite. Fulks [13] (see also [35]) proved that

$$\begin{aligned}&\frac{1}{\omega_m}\int_\Omega u(x+ra(x)^{1/2}s)H_{m-1}(ds)-u(x)\\ &=\frac{r^2}{2m}Lu(x)+O(r^2)\end{aligned} \tag{2.31}$$

as $r\to 0$, u being any twice continuously differentiable function. Here, $a(x)^{1/2}$ denotes the positive square root of the matrix $a(x)$ and H_{m-1} the $(m-1)$-dimensional measure. From (2.31) it is convenient to pass to the mean over an ellipsoid. Multiplying (2.31) by mr^{m-1} and integrating with respect to r, we obtain, using an obvious change of variables in the m-fold repeated integral

$$\begin{aligned}&\frac{1}{meas\ \mathcal{E}(x,r)}\int\limits_{\mathcal{E}(x,r)} u(y)dy - u(x)\\&= \frac{r^2}{2(m+2)}Lu(x) + O(r^2),\end{aligned} \tag{2.32}$$

where $\mathcal{E}(x,r)$ is the ellipsoid

$$\mathcal{E}(x,r) = \Big\{y \in \mathbb{R}^m :\ (a^{-1}(x)(y-x), y-x) < r^2\Big\},$$

with the center at x and axes on the eigenvectors of the matrix $a(x)$, while

$$meas\ \mathcal{E}(x,r) = (\omega_m/m)r^m[\det a(x)]^{1/2}.$$

Note that the boundaries $\partial\mathcal{E}(x,r)$ of the ellipsoids $\mathcal{E}(x,r)$ are the level surfaces of the Levi function

$$\mathcal{L}(x,y) = \frac{1}{(m-2)\omega_m[\det a(x)]^{1/2}}\Big(a^{-1}(x)(y-x), y-x\Big)^{1-m/2},$$

the fundamental solution of the constant elliptic operator L, x fixed.

Note that the same approach was used in [9] when constructing the Random Walk on ellipsoids.

Remark 2.4.

Let $C = (c_{jk})$ be a symmetric $n \times n$-matrtix and $r \neq 0$, and let

$$\sigma_x^{r,C}(u) = \int\limits_{\mathcal{E}_r} u(x+y)do(y),$$

where the integration is taken over the surface of the ellipsoid

$$\mathcal{E}_r = \Big\{y :\ \sum c_{jk}y_jy_k = r^2\Big\}.$$

It can be shown (e.g., see [17]) that if u is the solution to the following equation with constant matrix of coefficients $A = (a_{jk})$

$$\sum_{j,k=1}^{n} a_{jk}\frac{\partial^2 u}{\partial x_j \partial x_k} = 0,$$

then

$$\sigma_x^{r,C}(u) = \sigma_x^{r,C+tA}(u)$$

is true for all t. For the Laplace equation it means that the integrals of harmonic functions over the confocal ellipsoids do not depend on the parameter t (if the function is considered in a domain G, then t is such that the convex hull of $\operatorname{supp} \sigma_x^{r,C+tA}(u)$ lies in G).

2.4. Parabolic equations

2.4.1. Heat equation

Let Q be a domain in $\mathbb{R}^{n+1}$ whose points will be denoted by $(x,t) = (x_1, x_2, \ldots, x_n, t)$. We deal in this section with the heat equation

$$\frac{\partial u}{\partial t} = \Delta u(x,t) + f(x,t), \quad (x,t) \in Q. \tag{2.33}$$

Let

$$Z(x,t) = \theta(t)[4\pi t]^{-n/2} \exp\left\{\frac{|x|^2}{4t}\right\}$$

be the fundamental solution to (2.33), where $\theta(t)$ is the Heaviside step function: $\theta(t) = 0$ if $t \leq 0$ and $\theta(t) = 1$ if $t > 0$. Introduce the function

$$Z^{(\alpha)} = Z(x,t) - (4\pi\alpha)^{-n/2},$$

depending on a positive parameter α and define a family of domains $B_\alpha(x,t)$:

$$B_\alpha(x,t) = \left\{(x',t') \in \mathbb{R}^{n+1} : \ t' < t, \ Z^{(\alpha)}(x-x', t-t') > 0\right\} \tag{2.34}$$

with the boundaries

$$\partial B_\alpha(x,t) = \left\{(x',t') \in \mathbb{R}^{n+1} : \ t' \leq t, \ Z^{(\alpha)}(x-x', t-t') = 0\right\}. \tag{2.35}$$

We define also a family of domains $B_\alpha^{(\beta)}(x,t)$ depending on the parameter β, $0 < \beta \leq \alpha$:

$$B_\alpha^{(\beta)}(x,t) = B_\alpha(x,t) \cap \mathbb{R}^n \times (t-\beta, t), \tag{2.36}$$

and let

$$\partial B_\alpha^{(\beta)}(x,t) = \partial B_\alpha(x,t) \cap \mathbb{R}^n \times [t-\beta, t]. \tag{2.37}$$

We call the domain $B_\alpha(x,t)$ a balloid, and the set $\partial B_\alpha(x,t)$, a spheroid of radius α with the center at (x,t). The domains $B_\alpha^{(\beta)}(x,t)$, $\partial B_\alpha^{(\beta)}(x,t)$ are called truncated balloid and truncated spheroid, respectively.

We now obtain a mean value theorem where the solution of (2.33) at the point (x,t) is expressed in terms of its values integrated over a spheroid or a truncated spheroid at the center at (x,t). Then by averaging we derive the relations where the solution of (2.33) at the point (x,t) is expressed in terms of its values integrated over the spheroid and balloid at the center at (x,t).

Note that the balloid $B_\alpha(x,t)$ is situated between the planes $t' = t$ and $t' = t - \alpha$. The intersection of $B_\alpha(x,t)$ and the plane $t' = t - \tau$ $(0 < \tau < \alpha)$ is an n-dimensional ball $B(x, R(\tau))$ where

$$R(\tau) = \Big(2\tau n \ln(\alpha/\tau)\Big)^{1/2}.$$

The maximum of the radius $R(\tau)$ is attained at $\tau = \alpha/e$; its value is $R_{\max} = (2n\alpha/e)^{1/2}$. Hence,

$$B_\alpha(x,t) \subset B(x, R_{\max}) \times (t-\alpha, t), \tag{2.38}$$

and

$$\partial B_\alpha^{(\beta)}(x,t) \subset B(x, R_{\max}) \times (t-\beta, t). \tag{2.39}$$

The relation (2.38) shows that the balloid $B_\alpha(x,t)$ tends to the point (x,t) as $\alpha \to 0$. Therefore, for each point $(x,t) \in Q$, there exists α such that

$$B_\alpha(x,t) \subset Q. \tag{2.40}$$

Thus let $(x,t) \in Q$, and we suppose that α is chosen so that the condition (2.40) is satisfied. Obviously,

$$\frac{\partial}{\partial t'} Z^{(\alpha)}(x-x', t-t') + \Delta_{x'} Z^{(\alpha)}(x-x', t-t') = 0, \quad t < t', \tag{2.41}$$

where Δ_y is the Laplace operator (differentiation with respect to the variable y). From (2.33) and (2.41) we get

$$\begin{aligned}&Z^{(\alpha)}(x-x',t-t')\Delta_{x'}u(x',t')-u(x',t')\Delta_{x'}Z^{(\alpha)}(x-x',t-t')\\&=\frac{\partial}{\partial t'}\Big\{Z^{(\alpha)}(x-x',t-t')u(x',t')\Big\}-Z^{(\alpha)}(x-x',t-t')f(x',t').\end{aligned}\tag{2.42}$$

Integrating this equality in the ball $B(x,R(t-t'))$ we get by the Green formula

$$\begin{aligned}&\int\limits_{B(x,R(t-t'))}\frac{\partial}{\partial t'}\Big\{Z^{(\alpha)}(x-x',t-t')u(x',t')\Big\}dx'\\&=\int\limits_{\partial B(x,R(t-t'))}\Big\{Z^{(\alpha)}\frac{\partial u}{\partial n_{x'}}-u(x',t')\frac{\partial}{\partial n_{x'}}Z^{(\alpha)}\Big\}dS_{x'}\\&+\int\limits_{B(x,R(t-t'))}Z^{(\alpha)}(x-x',t-t')f(x',t')\Big\}dx',\end{aligned}\tag{2.43}$$

where $Z^{(\alpha)}=Z^{(\alpha)}(x-x',t-t')$, $n_{x'}$ is the exterior normal vector at x', and $dS_{x'}$ is the surface element of the sphere $\partial B(x,R(t-t'))$ at the point x'. Notice that $Z^{(\alpha)}(x-x',t-t')=0$ for $x'\in\partial B(x,R(t-t'))$ (by definition); hence we can change the order of the operator $\frac{\partial}{\partial t'}$ and the integral in the left-hand side of (2.43). Thus we get from (2.43)

$$\begin{aligned}&\frac{\partial}{\partial t'}\Big\{\int\limits_{B(x,R(t-t'))}Z^{(\alpha)}u(x',t')dx'\Big\}\\&=\int\limits_{\partial B(x,R(t-t'))}-\frac{\partial Z^{(\alpha)}}{\partial n_{x'}}u(x',t')dS_{x'}+\int\limits_{B(x,R(t-t'))}Z^{(\alpha)}f(x',t')dx'.\end{aligned}\tag{2.44}$$

We now integrate (2.44) with respect to t' over $[t-\beta,t]$, $(0<\beta\le\alpha)$:

$$\lim_{t'\to t}\int\limits_{B(x,R(t-t'))} Z^{(\alpha)}(x-x',t-t')u(x',t')dx'$$
$$-\int\limits_{B(x,R(\beta))} Z^{(\alpha)}(x-x',\beta)u(x',t-\beta)dx'$$
$$=\int\limits_{\partial B_\alpha^{(\beta)}(x,t)} -\frac{\partial Z^{(\alpha)}}{\partial n_{x'}}(x-x',t-t')u(x',t')dS_{x'}dt' \tag{2.45}$$
$$+\int\limits_{B_\alpha^{(\beta)}(x,t)} Z^{(\alpha)}(x-x',t-t')f(x',t')dx'dt'.$$

We prove that

$$\lim_{t'\to t}\int\limits_{B(x,R(t-t'))} Z^{(\alpha)}(x-x',t-t')u(x',t')dx' = u(x,t). \tag{2.46}$$

Indeed, let

$$\gamma(\rho,\tau) = (4\pi\tau)^{-n/2}\exp\left\{-\frac{\rho^2}{4\tau}\right\} - (4\pi\alpha)^{-n/2}$$

and let $\tau = t - t'$. Then

$$\int\limits_{B(x,R(\tau))} Z^{(\alpha)}(x-x',\tau)u(x',t-\tau)dx'$$
$$=\int\limits_0^{R(\tau)} \rho^{n-1}d\rho \int\limits_{S(0,1)} \gamma(\rho,\tau)u(x+\rho\omega,t-\tau)\,d\Omega(\omega)$$
$$=\int\limits_0^{R(\tau)} \rho^{n-1}d\rho \int\limits_{S(0,1)} \gamma(\rho,\tau)u(x,t-\tau)\,d\Omega(\omega) \tag{2.47}$$
$$+\int\limits_0^{R(\tau)} \rho^{n-1}d\rho \int\limits_{S(0,1)} \gamma(\rho,\tau)[u(x+\rho\omega,t-\tau)-u(x,t-\tau)]\,d\Omega(\omega),$$

where $d\Omega(\omega)$ is the surface element of the unit sphere $S(0,1)$.

Now,

$$\int_0^{R(\tau)} \rho^{n-1}\gamma(\rho,\tau)\omega_n d\rho$$

$$= \int_0^{R(\tau)} \frac{2\rho^{n-1}}{\Gamma(n/2)}\Big\{(4\tau)^{-n/2}\exp\{-\rho^2/4\tau\} - (4\alpha)^{-n/2}\Big\}d\rho \tag{2.48}$$

$$= \int_0^{\frac{n}{2}\ln\frac{\alpha}{\tau}} \frac{r^{n/2-1}}{\Gamma(n/2)}\Big(e^{-r} - \Big(\frac{\tau}{\alpha}\Big)^{n/2}\Big)dr \to 1$$

as $\tau \to 0$.

We used here the fact that $\omega_m = 2\pi^{n/2}/\Gamma(n/2)$. Analogously, we can prove that

$$\int_0^{R(\tau)} \rho^n\gamma(\rho,\tau)d\rho \to 0 \tag{2.49}$$

as $\tau \to 0$.

We get from (2.47) by (2.48) and (2.49) the desired relation (2.46). Thus, from (2.45) and in view of (2.46), we get the following mean value relation.

Theorem 2.7.

If the parameters α and β, $0 < \beta \le \alpha$, are chosen so that $B_\alpha^{(\beta)}(x,t) \subset Q$, then the regular solutions to the heat equation (2.33) satisfy the relation

$$\begin{aligned} u(x,t) = &\int_{B(x,R(\beta))} Z^{(\alpha)}(x-x',\beta)u(x',t-\beta)dx' \\ &+ \int_{\partial B_\alpha^{(\beta)}(x,t)} \Big(-\frac{\partial Z^{(\alpha)}}{\partial n_{x'}}\Big)u(x',t')dS_{x'}dt' \\ &+ \int_{B_\alpha^{(\beta)}(x,t)} Z^{(\alpha)}(x-x',t-t')f(x',t')dx'dt'. \end{aligned} \tag{2.50}$$

An important particular case of (2.50) is $\beta = \alpha$:

$$u(x,t) = \int\limits_{\partial B_\alpha(x,t)} -\frac{\partial Z^{(\alpha)}}{\partial n_{x'}}(x-x',t-t')u(x',t')dS_{x'}dt' + \int\limits_{B_\alpha(x,t)} Z^{(\alpha)}(x-x',t-t')f(x',t')dx'dt' \tag{2.51}$$

which was derived in [22]. Further mean value relations can be obtained by integration of (2.50) with respect to β over the interval $[0,\alpha]$. This yields

Theorem 2.8.

Under the assumptions of the previous theorem, the following mean value relation holds

$$u(x,t) = \int\limits_{\partial B_\alpha(x,t)} \Big(1-\frac{t-t'}{\alpha}\Big)\Big(-\frac{\partial Z^{(\alpha)}}{\partial n_{x'}}(x-x',t-t')\Big)u(x',t')dS_{x'}dt' + \frac{1}{\alpha}\int\limits_{B_\alpha(x,t)} Z^{(\alpha)}(x-x',t-t')u(x',t')dx'dt' + F_\alpha(x,t), \tag{2.52}$$

where

$$F_\alpha(x,t) = \int\limits_{B_\alpha(x,t)} \Big(1-\frac{t-t'}{\alpha}\Big)Z^{(\alpha)}(x-x',t-t')f(x',t')dx'dt'. \tag{2.53}$$

Here we used the relations

$$\int_0^\alpha d\beta \int\limits_{B(x,R(\beta))} Z^{(\alpha)}(x-x',\beta)u(x',t-\beta)\,dx' = \int\limits_{B_\alpha(x,t)} Z^{(\alpha)}(x-x',t-t')u(x',t')dx'dt',$$

$$\int_0^\alpha d\beta \int_{\partial B_\alpha^{(\beta)}(x,t)} -\frac{\partial Z^{(\alpha)}}{\partial n_{x'}}(x-x',t-t')u(x',t')\,dS_{x'}dt'$$
$$= \int_{\partial B_\alpha(x,t)} (\alpha-(t-t'))\Big(-\frac{\partial Z^{(\alpha)}}{\partial n_{x'}}\Big)u(x',t')dS_{x'}dt',$$

$$\int_0^\alpha d\beta \int_{B_\alpha^{(\beta)}(x,t)} Z^{(\alpha)}(x-x',t-t')f(x',t')\,dx'dt'$$
$$= \int_{B_\alpha(x,t)} (\alpha-(t-t'))Z^{(\alpha)}(x-x',t-t')f(x',t')dx'dt'.$$

Remark 2.5.

Let us consider the equation

$$\frac{\partial u}{\partial t} - \Delta u(x,t) + c(x,t)u = f(x,t), \quad (x,t)\in Q \tag{2.54}$$

with $0 \leq c(x,t) \leq c_0$. Suppose that the parameter α is chosen so that $B_\alpha(x,t) \subset Q$. Then the regular solutions to the equation (2.54) satisfy the following mean value relation

$$u(x,t) = \int_{\partial B_\alpha(x,t)} \Big(1-\frac{t-t'}{\alpha}\Big)\Big(-\frac{\partial Z^{(\alpha)}}{\partial n_{x'}}(x-x',t-t')\Big)u(x',t')dS_{x'}dt'$$
$$+\frac{1}{\alpha}\int_{B_\alpha(x,t)} [1-(\alpha-(t-t'))c(x',t')]Z^{(\alpha)}(x-x',t-t')u(x',t')dx'dt'$$
$$+F_\alpha(x,t),$$

where $F_\alpha(x,t)$ is defined in (2.53). The proof follows from (2.52).

2.4.2. Parabolic equations with variable coefficients

In this section we generalize the results obtained for the heat equation to the following equation

$$\frac{\partial u}{\partial t} - \mu\Delta u(x,t) + \sum_{i=1}^{n} b_i(x,t)\frac{\partial u}{\partial x_i} + c(x,t)u = f(x,t), \quad (x,t) \in Q. \quad (2.55)$$

We assume that the coefficients $b_i(x,t)$ and their derivatives $\partial b_i/\partial x_i$ are continuous and bounded on Q, the coefficient $c(x,t)$ and the function $f(x,t)$ are also continuous and bounded on Q. We employ in this section the summation convention (summation over the repeated index).

Let

$$Z(x,t;x',t') = \frac{\theta(t-t')}{[4\pi\mu(t-t')]^{n/2}} \exp\left\{-\frac{|x-x'-(t-t')b(x,t)|^2}{4\mu(t-t')}\right\}, \quad (2.56)$$

where $\theta(t)$ is the Heaviside step function, $b(x,t) = (b_1(x,t),\dots,b_n(x,t))$.

The function $Z(x,t;x't')$ solves the equation

$$-\frac{\partial Z}{\partial t'} - \mu\Delta_{x'}Z(x,t;x',t') - b_i(x,t)\frac{\partial Z}{\partial x_i'}(x,t;x',t') = 0, \quad t' < t. \quad (2.57)$$

Introduce the function

$$Z^{(\alpha)}(x,t;x',t') = Z(x,t;x',t') - (4\pi\mu\alpha)^{-n/2}. \quad (2.58)$$

Obviously,

$$-\frac{\partial Z^{(\alpha)}}{\partial t'} - \mu\Delta_{x'}Z^{(\alpha)}(x,t;x',t') - b_i(x,t)\frac{\partial Z^{(\alpha)}}{\partial x_i'}(x,t;x',t') = 0, \quad (2.59)$$

for $t' < t$.

As in the previous section, we define the families of domains:

$$B_\alpha(x,t) = \left\{(x',t') \in \mathbb{R}^{n+1};\ t' < t,\ Z^{(\alpha)}(x,t;x',t') > 0\right\}$$

is a balloid,

$$\partial B_\alpha(x,t) = \left\{(x',t') \in \mathbb{R}^{n+1};\ t' \le t,\ Z^{(\alpha)}(x,t;x',t') = 0\right\}$$

is a spheroid, and the following two-parametric sets

$$B_\alpha^{(\beta)}(x,t) = \left\{(x',t') \in \mathbb{R}^{n+1};\ t-\beta < t' < t,\ Z^{(\alpha)}(x,t;x',t') > 0\right\},$$

$$\partial B_\alpha^{(\beta)}(x,t) = \Big\{(x',t') \in \mathbb{R}^{n+1};\ t-\beta \le t' \le t,\ Z^{(\alpha)}(x,t;x',t') = 0\Big\}.$$

Note that the structure of these balloids is somewhat more complicated: the intersection of $B_\alpha(x,t)$ with the plane $t' = t-\tau$, $(0 < \tau < \alpha)$ is the ball in $\mathbb{R}^n$ centered at the point $x(\tau) = x - \tau b(x,t)$ with the radius $R(\tau) = (2\tau n\mu \ln(\alpha/\tau))^{1/2}$. It is clear that for arbitrary point $(x,t) \in Q$ there is a value of α such that $B_\alpha^{(\beta)}(x,t) \subset Q$.

Theorem 2.9.

Assume that the parameters α and β, $0 < \beta \le \alpha$, are chosen so that $B_\alpha^{(\beta)}(x,t) \subset Q$. Then the regular solutions to the equation (2.55) satisfy the following mean value relation

$$\begin{aligned} u(x,t) = & \int\limits_{B(x(\beta),R(\beta))} Z^{(\alpha)}(x,t;x',t-\beta)u(x',t-\beta)dx' \\ + & \int\limits_{\partial B_\alpha^{(\beta)}(x,t)} \Big(-\mu\frac{\partial Z^{(\alpha)}}{\partial n_{x'}}\Big)u(x',t')dS_{x'}dt' \\ + & \int\limits_{B_\alpha^{(\beta)}(x,t)} k(x,t;x',t')Z^{(\alpha)}(x,t;x',t')u(x',t')dx'dt' \\ + & \int\limits_{B_\alpha^{(\beta)}(x,t)} Z^{(\alpha)}(x,t;x',t')f(x',t')dx'dt', \end{aligned}$$

where

$$\begin{aligned} k(x,t;x',t') = & \frac{b_i(x,t)-b_i(x',t')}{2\mu(t-t')}\Big[x_i' - x_i - (t-t')b_i(x,t)\Big] \\ & + \frac{\partial b_i}{\partial x_i'}(x',t') - c(x',t'). \end{aligned} \tag{2.60}$$

Proof. From (2.55) and (2.59) we find

$$
\begin{aligned}
&\mu\Big\{Z^{(\alpha)}(x,t;x',t')\Delta_{x'}u(x',t') - u(x',t')\Delta_{x'}Z^{(\alpha)}(x,t;x',t')\Big\} \\
&= \frac{\partial}{\partial t'}\Big\{Z^{(\alpha)}(x,t;x',t')u(x',t')\Big\} + \frac{\partial}{\partial x_i'}\Big\{b_i(x',t')u(x',t')Z^{(\alpha)}\Big\} \\
&+\Big(b_i(x,t) - b_i(x',t')\Big)\frac{\partial Z^{(\alpha)}}{\partial x_i'}(x,t;x',t')u(x',t') \\
&+\Big[c(x',t') - \frac{\partial b_i}{\partial x_i'}(x',t')\Big]Z^{(\alpha)}(x,t;x',t')u(x',t') \\
&+f(x',t')Z^{(\alpha)}(x,t;x',t').
\end{aligned}
\tag{2.61}
$$

Integrating this equality, first, with respect to x over the ball $B(x(t-t'), R(t-t'))$, then with respect to t' over $(t-\beta, t)$, and using the Green formulae (keeping in mind $Z^{(\alpha)}(x,t;x',t') \equiv 0$ for $(x',t') \in \partial B_\alpha^{(\beta)}(x,t)$), we get the desired result. □

2.4.3. Expansion of the parabolic means

It is possible to derive expansions of the means which generalize the expansions (2.7).

Let $x \in \mathbb{R}^n$, $t \in \mathbb{R}^1$, and let $\alpha = (\alpha^{ij})$ be a constant positive definite symmetric $n \times n$ matrix whose entries α^{ij} are real values. In $\mathbb{R}^{n+1} = \{(x,t) : t \in \mathbb{R}^1\}$ we consider the parabolic equation

$$\frac{\partial u}{\partial t} - Lu = 0 \tag{2.62}$$

where the operator L is defined by

$$Lu \equiv \sum_{i,j=1}^{n} \alpha^{ij}\frac{\partial^2 u}{\partial x^i \partial x^j}\ .$$

Let us consider a set B_r in $\mathbb{R}^{n+1}$ which we call a balloid of radius r:

$$B_r = \Big\{(y,\tau) :\ y^T\alpha^{-1}y < 2(n+m)\tau\ln(r^2/\tau)\Big\}, \tag{2.63}$$

where m, r are positive real numbers, the matrix α defined above. It is clear that in (2.63) $0 < \tau < r^2$, and the vector y is so that

$$y^T\alpha^{-1}y < [2(n+m)/e]r^2,$$

i.e., the horizontal diameter of the balloid is a quantity of order r. As $r \to 0$, the spheroids shrink monotonically to the point $(0,0)$. Let us consider the following parabolic mean

$$(M_r u)(x,t) = \iint_{B_r} u(x+y, t-\tau)K(y,\tau)dy d\tau, \qquad (2.64)$$

where

$$K(y,\tau) = \frac{n+m}{|\alpha|^{1/2}\pi^{n/2}2^{n+m-1}\Gamma(m/2)r^{n+m}} \times \ln(r^2/\tau)[2(n+m)\tau\ln(r^2/\tau) - y^T\alpha^{-1}y]^{m/2-1}\ . \qquad (2.65)$$

Let $p \geq 0$ be an integer. We say that $u(x,t) \in C^{2p,p}(G)$ if all the derivatives

$$\frac{\partial}{\partial x_{i_1}}\frac{\partial}{\partial x_{i_2}}\cdots\frac{\partial}{\partial x_{i_k}}\left(\frac{\partial}{\partial t}\right)^l u(x,t)$$

exist and are continuous in G for all pairs (k,l) of integers satisfying the conditions

$$k \geq 0,\ l \geq 0,\ k+2l \leq 2p\ .$$

Theorem 2.10.

Assume that $u(x,t) \in C^{2p+2,p+1}(\bar{G})$, $p \geq 0$. Then for all inner points $(x,t) \in G$ and for all sufficiently small values r (such that $0 < r < d(x,t)$) the following expansion is true

$$(M_r u)(x,t) = \sum_{i=0}^{p} r^{2i}L_i u(x,t) + Q_p(r)u(x,t), \qquad (2.66)$$

where the differential operators are defined by

$$L_i = \frac{(-1)^i}{i!}\left(\frac{n+m}{n+m+2i}\right)^{(n+m)/2+1}\left(\frac{\partial}{\partial t} - \frac{n+m}{n+m+2i}L\right)^{i-1} \times \left(\frac{\partial}{\partial t} - L\right) \qquad (2.67)$$

for $i \geq 1$, $L_0 = 1$, and $Q_p(r)u(x,t) = O(r^{2p+1})$ as $r \to 0$.

If $u(x,t) \in C^\infty(G)$ satisfies the condition

$$\lim_{p\to\infty} Q_p(r)u(x,t) = 0, \tag{2.68}$$

then we can pass to the limit as $p \to \infty$. Thus for the solutions to the equation $\frac{\partial u}{\partial t} = Lu(x,t)$ satisfying (2.68), we have the following mean value relation

$$u(x,t) = (M_r u)(x,t) \tag{2.69}$$

for all $(x,t) \in G$ and all r, $0 < r < r(x,t)$. It is not difficult to show that this property is true for each solution of $\frac{\partial u}{\partial t} = Lu$ from the class $C^{2,1}$.

Conversely, if a function $u \in C^{4,2}$ satisfies (2.69), then we get from the expansion that

$$0 = r^2 Lu(x,t) + O(r^3),$$

or $0 = Lu(x,t) + O(r)$. Then we get by $r \to 0$: $Lu(x,t) = 0$. This implies the weak mean value theorem. A strong variant of this statement can be proved using the maximum principle according to the scheme given for the Laplace equation.

Note that different expansions of the type (2.66) can be derived. For instance, choosing

$$\begin{aligned} K(y,\tau) &= \frac{1}{|\alpha|^{1/2}\pi^{n/2}2^{n+m}\Gamma(m/2)r^{n+m^2+2m}} \\ &\times \tau^{(m^2+m)/2-1}[2(n+m^2+2m)\tau\ln(r^2/\tau) - y^T\alpha^{-1}y]^{m/2-1} \\ &\times [(m+2)(2(n+m^2+2m)\tau\ln(r^2/\tau) - y^T\alpha^{-1}y) + y^T\alpha^{-1}y] \end{aligned} \tag{2.70}$$

and

$$B_r = \{(y,\tau):\ y^T\alpha^{-1}y < 2(n+m^2+2m)\tau\ln(r^2/\tau)\}$$

we come to an expansion of the type (2.66) with

$$L_i = \frac{(-1)^i}{i!}[g(n,m)]^{(n+m)/2+1}\left(\frac{\partial}{\partial t} - g(n,m)L\right)^{i-1}\left(\frac{\partial}{\partial t} - L\right), \quad (2.71)$$

where

$$g(n,m) = \frac{n + m^2 + 2m}{n + m^2 + 2m + 2i} .$$

Note that both expansions with (2.66) and (2.71) result, when $m \to +0$, in the expansions with

$$L_i = \frac{(-1)^i}{i!}\left(\frac{n}{n+2i}\right)^{n/2+1} \times \left(\frac{\partial}{\partial t} - \frac{n}{n+2i}L\right)^{i-1}\left(\frac{\partial}{\partial t} - L\right). \quad (2.72)$$

This is the expansion of the average of the function $u(x,t)$ over the surfaces (spheroids)

$$\{(y,\tau):\ Z(x,t,y,\tau) = \pi^{-n/2}|\alpha|^{-1/2}r^{-n}\} .$$

Chapter 3

High-order elliptic equations

3.1. Balayage operator

In this section we give a general scheme of construction of the spherical mean value relations for high-order elliptic equations (see [37] and [38]).

Let X be a locally compact Hausdorf space, and let $C_0(X)$ be a space of continuous functions $f : X \to \mathbb{R}$ with a compact support. We denote by $R(X)$ the space of Radon measures $\mu : C_0(X) \to \mathbb{R}$.

Let $G \subset \mathbb{R}^n$ be a bounded domain with the boundary $\partial G = \Gamma$. By $C^k(\bar{G})$ we denote the space of continuous on $\bar{G}$ functions $u(x)$ such that the derivatives $D^\alpha u$ exist in G for all $|\alpha| \leq k$ and admit continuous continuation to $\bar{G}$. Let $C_0^\infty(G)$ be a space of infinitely differentiable functions with a compact support in G. Introduce a vector measure

$$\mu = (\mu^\alpha)_{|\alpha|\leq k} \in \bigotimes_{|\alpha|\leq m-1} R(\mathbb{R}^n), \quad \operatorname{supp}\mu = \bigcup_{|\alpha|\leq k} \operatorname{supp}\mu^\alpha \subset G\ .$$

Let us consider an arbitrary elliptic operator in $\mathbb{R}^n$ of order $2m$ with real-valued coefficients $a_\alpha \in C^\infty(\mathbb{R}^n)$:

$$L(x, D) = \sum_{|\alpha|\leq 2m} a_\alpha(x) D^\alpha \tag{3.1}$$

and m differential operators $B_j(x, D)$ with coefficients from $C_0^\infty(\mathbb{R}^n)$ such that $\{\text{the order of } B_j\} = m_j \leq 2m - 1$.

Let us consider the boundary value problem

$$\begin{aligned} L(x, D)u &= 0, \quad x \in G\ , \\ B_j(x, D)u &= g_j, \quad j = 1, 2, \dots, m, \quad x \in G\ . \end{aligned} \tag{3.2}$$

We will assume that the functions g_j, the operators B_j and the boundary Γ satisfy the conditions which guarantee that the problem (3.2) has a unique solution (concerning these conditions, see, for example, [50]). We assume also that the generalized maximum principle is satisfied:

$$\sup_{x\in G}\sum_{k=1}^{s}|Q_k(x,D)u(x)| \le c\sum_{j=1}^{m}\|B_j(x,D)u(x)\Big|_{\Gamma}\|_{C^{l-m_j}(\Gamma)}\ , \tag{3.3}$$

where Q_k are differential operators of order $\le l$ with smooth coefficients (they are related on Γ to the operators $B_j(x,D)$; see [50]). In particular, for the Dirichlet problem

$$L(x,D)u = 0,\quad x\in G\ ,$$
$$D^\alpha u\Big|_\Gamma = D^\alpha v\Big|_\Gamma,\quad |\alpha|\le m-1,\quad v\in C^{m-1}(\mathbb{R}^n)$$

it reads

$$\sup_{x\in G}\sum_{|\alpha|\le m-1}|D^\alpha u(x)| \le c\sup_{\Gamma}\sum_{|\alpha|\le m-1}|D^\alpha u|\ .$$

We denote by Π the balayage operator [50] for the problem (3.2):

$$\Pi:\mu\longrightarrow\nu,\quad \operatorname{supp}\mu\subset\bar G,\quad \operatorname{supp}\nu\subset\Gamma\ .$$

Let us introduce a functional space

$$H(\bar G)=\{h=(D^\alpha u)_{|\alpha|\le m-1}: u\in C^{m-1}(\bar G)\cap C^{2m}(G),\ L(x,D)u=0\,\text{in}\,G\}.$$

The following balayage principle is true under the above assumptions and (3.3): for an arbitrary measure $\mu=(\mu^\alpha)_{|\alpha|\le m-1}$, $\operatorname{supp}\mu\subseteq G$ there exists a measure $\nu=(\nu^\alpha)_{|\alpha|\le m-1}$, $\operatorname{supp}\nu\subseteq\Gamma$ such that

$$\mu(h)=\nu(h)\ ,\quad \text{for all } h\in H(\bar G)\ . \tag{3.4}$$

Let $B(x_0,r)$ be a ball, and let $\Pi B(x_0,r)$ be the balayage operator for the problem $Lu=0$ in the ball $B(x_0,r)\subset G$ where r is sufficiently small. Let $\sigma_{x_0}^r$ be a measure uniformly distributed on the sphere $S(x_0,r)$:

$$\sigma_{x_0}^r(u)=c(r)\int\limits_{S(x_0,r)}u(y)\,dS(y)\ ,$$

where $c(r)$ is a normalizing factor. Throughout this chapter we shall derive mainly spherical mean value relations generated by the measure $\sigma_{x_0}^r(u)$.

Let δ_x be a unit mass concentrated at the point x, and let δ_{x_0,α_0} be a vector measure whose components are equal to zero for $\alpha \neq \alpha_0$, while they are equal to δ_{x_0} if $\alpha = \alpha_0$.

Now, let us take a sphere $S(x_0, r) \subset G$. Then, under the same assumptions as stipulated above when formulating (3.4), there exists a vector measure $\nu^j = (\nu^{j_\beta})_{|\beta| \leq l-m_j}$, supp $\nu^{j_\beta} \subset S(x_0, r)$ such that the solutions to the problem (3.2) satisfy the system

$$Q_k u(x) = \sum_{j=1}^{s} \sum_{|\beta| \leq l-m_j} Q_{j_\beta} u(x') d\nu_x^{j_\beta}(x') . \qquad (3.5)$$

Indeed from (3.4) it follows that

$$\sum_{k=1}^{s} \int_{\bar{G}} Q_k(x, D) u(x) d\mu^k(x) = \sum_{j=1}^{m} \nu^j(B_j(x, D) u(x)) .$$

Choosing here the measures μ^k equal to $\delta_{x_0,k} = \Pi B(x_0, r) \delta_{x_0,k}$ and the representation [50]

$$\nu^j(v) = \sum_{|\beta| \leq l-m_j} \int_{\Gamma} D^\beta v(x) d\nu^{j_\beta}(x)$$

we get (3.5).

This is a general statement which ensures that under the formulated assumptions it is possible to derive spherical mean value relations of the types (3.5). Concrete relations will be derived in this chapter for a polyharmonic and metaharmonic equations.

3.2. The biharmonic equation

For the high order PDEs it is possible to derive spherical mean value relations of different type. For instance, in [34] (see also [30]) a series of spherical mean value relations for polyharmonic equation is derived which relates the spherical mean and the powers of the Laplacian at a fixed point. However, our objective is to construct a mean value relation which uniquely characterizes the relevant boundary value problem. We will see that a vector generalization of the converse spherical mean value theorem presented above for the harmonic functions can be formulated which provides a system of integral equations equivalent to the original boundary value problem under discussion. In this section we deal with the biharmonic equation

$$\Delta\Delta u(x) = 0, \ x \in G , \tag{3.6}$$

which governs (in $2D$) the bending of a thin elastic plate. We will consider two cases:

(a) simply supported boundary of the plane

$$u\Big|_{\Gamma} = g_0, \quad \left\{\Delta u - \frac{1-\sigma}{\rho}\frac{\partial u}{\partial n}\right\}\Big|_{\Gamma} = g_1 , \tag{3.7}$$

and

(b) rigid fixing of the boundary

$$u\Big|_{\Gamma} = g_0, \quad \frac{\partial u}{\partial n}\Big|_{\Gamma} = g_1. \tag{3.8}$$

Here σ (=const) is the Poisson's ratio of the plate, n is the exterior normal vector to the boundary Γ and ρ is the curvature of the boundary.

3.2.1. Direct spherical mean value relation

In this section we treat the case (a) for the plates with $\rho = \infty$, so that the second boundary condition in (3.7) is $\Delta u|_\Gamma = g_1$.

We seek the regular solution $u \in C^4(G)\bigcap C^2(G\bigcup\Gamma)$ to the biharmonic equation

$$\Delta^2 u(x) = 0 \ , \ x \in G \subset \mathbb{R}^m . \tag{3.9}$$

Let us recall (see the previous chapter) that, for any analytic function v defined in $\mathbb{R}^m$, for any x and r such that $S(x,r) \subset G$, one has

$$\begin{aligned} N_r v(x) &= v(x) + \frac{r^2}{2m}\Delta v(x) \\ &+ \sum_{k=2}^{\infty}\left\{\frac{r^{2k}}{2^k k! m(m+2)\cdots(m+2k-2)}\,\Delta^k v(x)\right\} . \end{aligned} \tag{3.10}$$

Using the expansion (3.10) we get the following mean value relation :

$$u(x) = N_r u(x) - \frac{r^2}{2m} N_r(\Delta u)(x) \quad , \tag{3.11}$$

$$\Delta u(x) = N_r(\Delta u)(x) \; . \tag{3.12}$$

Of course, it is possible to derive this mean value relation using the Gauss formula, as in the case of the Laplace equation (see Chapter 2). However, we give in Section 3.2.2 a different derivation which provides the spherical mean value relation for an arbitrary point inside the sphere. This means actually that for this problem, the generalized Poisson formula for a ball is obtained.

Now we give the converse mean value relation which presents an integral formulation of the boundary value problem (3.9), (3.7) (we formulate it for an arbitrary biharmonic function in $G \subset \mathbb{R}^m$).

Proposition 3.1. Integral formulation.

Suppose the problem (3.9), (3.7) *is uniquely solvable. Assume that* v_1, v_2 *belong to* $\{C(G \bigcup \Gamma);\ \ v_1|_\Gamma = g_0,\ v_2|_\Gamma = g_1\}$ *and for each* $x \in G$ *there exists a sphere* $S(x,r) \subset \bar{G}$ *for which the following mean value relation holds :*

$$v_1 = N_r v_1 - \frac{r^2}{2m} N_r v_2 \quad ,$$

$$v_2 = N_r v_2 \quad .$$

Then $\Delta u = v_2$, *and the function* v_1 *is the unique solution to the problem* (3.9), (3.7).

Proof. Let us take the solution to (3.9), (3.7). It satisfies (3.11), (3.12); thus $u - v_1$ satisfies

$$u - v_1 = N_r(u - v_1) - \frac{r^2}{2m} N_r(\Delta u - v_2) \quad .$$

Using the results from the second section, we get that $\Delta v_2(x) = 0$ in G ; since Δu is harmonic, and $\Delta u|_\Gamma = v_2|_\Gamma = g_1$, we have $\Delta u \equiv v_2$ in G ; thus,

$$u - v_1 = N_r(u - v_1) \quad ,$$

which implies that the function $u - v_1$ is harmonic in G. Since $u|_\Gamma = v_1|_\Gamma = g_0$, we get $u \equiv v_1$ in G. □

3.2.2. The generalized Poisson formula

Let us consider a $2D$ domain, and let $S(x_0, r) = \partial K(x_0, r)$ be a circle of radius r centered at x_0. We use the notation

$$dS = r d\omega_2 = \omega_2 r d\sigma^r_{x_0} , \quad dx = t dt d\omega_2 , \quad \omega_2 = 2\pi .$$

Let α be the angle between the vectors $x - x_0$ and $y - x_0$, where y is the vector lying on the sphere, and let $t = |x - x_0|$.

Theorem 3.1.

The regular solution to the biharmonic equation satisfies the following spherical mean value relation

$$\begin{aligned} u(x) = {} & \frac{r^2 - t^2}{r\omega_2} \int\limits_{S(x_0,r)} \frac{u(y) dS_y}{|x - y|^2} \\ & + \frac{r^2 - t^2}{r\omega_2} \int\limits_{S(x_0,r)} \Big\{ \frac{r \cos^2 \alpha}{|x - x_0| \sin \alpha} \mathrm{arctg} \Big\{ \frac{|x - x_0| \sin \alpha}{r - |x - x_0| \cos \alpha} \\ & - \frac{1}{2} - \frac{r \cos \alpha}{2t} \ln \frac{|x - y|^2}{r^2} \Big\} \Delta u(y) dS_y , \end{aligned} \tag{3.13}$$

$$\Delta u(x) = \frac{r^2 - t^2}{r\omega_2} \int\limits_{S(x_0,r)} \frac{\Delta u(y) dS_y}{|x - y|^2} . \tag{3.14}$$

Proof. The biharmonic equation has the following general representation (e.g. see [55])

$$u(x) = (|x - x_0|^2 - r^2) v(x) + w(x) , \tag{3.15}$$

where $v(x)$, $w(x)$ are harmonic functions.

For the boundary points we have

$$w(y) = u(y) := g_0(y), \quad y \in S(x_0, r) .$$

The Poisson formula reads [4]

$$w(x) = \frac{r^2 - t^2}{r\omega_2} \int\limits_{S(x_0,r)} \frac{g_0(y)}{\rho^2} dS_y \ , \tag{3.16}$$

where $\rho = |x - y|$.

It remains to find the function v. From (3.15) we get

$$\begin{aligned}
\Delta u(x) &= \Delta\{(|x - x_0|^2 - r^2)v(x)\} + \Delta w(x) \\
&= v(x)\Delta(|x - x_0|^2 - r^2) + 2\frac{\partial v(x)}{\partial x_1}\frac{\partial}{\partial x_1}\left(|x - x_0|^2 - r^2\right) \\
&+ 2\frac{\partial v}{\partial x_2}\frac{\partial}{\partial x_2}\left(|x - x_0|^2 - r^2\right) \\
&+ (|x - x_0|^2 - r^2)\Delta v(x) + \Delta w(x) \ .
\end{aligned} \tag{3.17}$$

Let $x = (x_1, x_2)$, $x_0 = (x_1^0, x_2^0)$. It is easy to show that

$$\begin{aligned}
\Delta u(x) &= 4v(x) + 4(x_1 - x_1^0)\frac{\partial v(x)}{\partial x_1} + 4(x_2 - x_2^0)\frac{\partial v(x)}{\partial x_2} \\
&= 4v(x) + 4r\left[\frac{x_1 - x_1^0}{r}\frac{\partial v(x)}{\partial x_1} + \frac{x_2 - x_2^0}{r}\frac{\partial v(x)}{\partial x_2}\right] \ .
\end{aligned} \tag{3.18}$$

For x on the surface of the sphere

$$\frac{\partial v(x)}{\partial n} = \frac{x_1 - x_1^0}{r}\frac{\partial v(x)}{\partial x_1} + \frac{x_2 - x_2^0}{r}\frac{\partial v(x)}{\partial x_2} \ , \tag{3.19}$$

where n is the normal vector exterior to the surface $S(x, r)$.

Hence we get from (3.18) for the boundary points:

$$g_1(y) := \Delta u(y) = 4v(y) + 4r\frac{\partial v(y)}{\partial n} \ .$$

Thus we conclude that the function $v(x)$ is the unique solution of the following boundary value problem

$$\Delta v(x) = 0, \quad x \in K(x_0, r) \ , \tag{3.20}$$

$$\frac{\partial v}{\partial n} + \frac{1}{r}v(y) = \frac{g_1(y)}{4r}, \quad y \in S(x, r) \ . \tag{3.21}$$

To solve the problem (3.20), (3.21) we use the polar coordinates $x = \rho\exp(\mathrm{i}\varphi)$. Then we assume that there is an expansion

$$v(x) = v(\rho e^{i\varphi}) = \sum_{n=1}^{\infty} \frac{\rho^n (A_n \cos(n\varphi) + B_n \sin(n\varphi))}{r^{n-1}(n+1)} + A_0 \cdot \frac{r}{2}, \quad (3.22)$$

where A_n, B_n are the Fourier coefficients of the expansion of the function $-g_1(\rho e^{i\varphi})/4r$:

$$\begin{aligned} A_n &= -\frac{1}{4\pi r} \int_0^{2\pi} g_1(re^{i\theta}) \cos(n\theta)\, d\theta, \quad n = 0, 1, \ldots, \\ B_n &= -\frac{1}{4\pi r} \int_0^{2\pi} g_1(re^{i\theta}) \sin(n\theta)\, d\theta, \quad n = 0, 1, \ldots . \end{aligned} \quad (3.23)$$

Using

$$A_n \cos(n\varphi) + B_n \sin(n\varphi) = -\frac{1}{4\pi r} \int_0^{2\pi} g_1(re^{i\theta}) \cos[n(\theta - \varphi)]\, d\theta ,$$

$$A_0 = -\frac{1}{4\pi r} \int_0^{2\pi} g_1(re^{i\theta})\, d\theta ,$$

we get

$$v(\rho e^{i\varphi}) = -\frac{1}{4\pi r} \int_0^{2\pi} g_1(re^{i\theta}) \sum_{n=1}^{2\pi} \frac{\rho^n \cos[n(\theta - \varphi)]}{r^n(n+1)}\, d\theta. \quad (3.24)$$

Let $a = \rho/r$, and let

$$R(a, \theta) = \sum_{n=1}^{\infty} \frac{a^n \cos(n\theta)}{n+1} . \quad (3.25)$$

The function

$$R_1(t, \theta) = \frac{1}{2} + \sum_{n=1}^{\infty} t^n \cos(n\theta)$$

is uniformly convergent for $t^2 < 1$, and

$$R_1(t,\theta) = \frac{1}{2}\frac{1-t^2}{1-2t\cos\theta+t^2}\,.$$

The uniform convergence for $t < 1$ implies that it is possible to carry out the termwise integration which gives

$$R(a,\theta) = \frac{1}{a}\int_0^a R_1(t,\theta)dt - \frac{1}{2}\,. \tag{3.26}$$

Let us evaluate the integral

$$I = \frac{1}{a}\int_0^a R_1(t,\theta)dt = \frac{1}{2a}\int_0^a \frac{1-t^2}{1-2t\cos\theta+t^2}\,dt. \tag{3.27}$$

Using the known integrals [7]

$$\frac{1}{2a}\int_0^a \frac{dt}{1-2t\cos\theta+t^2} = \frac{1}{2a\sin\theta}\left[\operatorname{arctg}\frac{a-\cos\theta}{\sin\theta} - \operatorname{arctg}(-\operatorname{ctg}\theta)\right], \tag{3.28}$$

and

$$\frac{1}{2a}\int_0^a \frac{t^2\,dt}{1-2t\cos\theta+t^2} = \frac{1}{2} + \frac{\cos\theta}{2a}\ln|1-2a\cos\theta+a^2|$$
$$- \frac{\cos^2\theta - 1/2}{a\sin\theta}\left[\operatorname{arctg}\frac{a-\cos\theta}{\sin\theta} - \operatorname{arctg}(-\operatorname{ctg}\theta)\right]$$

we find

$$I = \frac{\cos^2\theta}{a\sin\theta}\left\{\operatorname{arctg}\frac{a-\cos\theta}{\sin\theta} - \operatorname{arctg}(-\operatorname{ctg}\theta)\right\}$$
$$- \frac{1}{2} - \frac{\cos\theta}{2a}\ln|1-2a\cos\theta+a^2|. \tag{3.29}$$

Let

$$\alpha = \operatorname{arctg}\frac{a-\cos\theta}{\sin\theta}, \quad \beta = \operatorname{arctg}(-\operatorname{ctg}\theta),$$

then

$$\operatorname{tg}\alpha = \frac{a-\cos\theta}{\sin\theta}, \quad \operatorname{tg}\beta = -\operatorname{ctg}\theta.$$

Simple transformations yield

$$\alpha - \beta = \operatorname{arctg}\Big\{a\sin\theta/(1 - a\cos\theta)\Big\}.$$

Hence,

$$I = \frac{\cos^2\theta}{a\sin\theta}\operatorname{arctg}\frac{a\sin\theta}{1 - a\cos\theta} - \frac{1}{2} - \frac{\cos\theta}{2a}\ln|1 - 2a\cos\theta + a^2|. \qquad (3.30)$$

We rewrite (3.24), using (3.25)-(3.27),

$$\begin{aligned} v(\rho e^{i\varphi}) = &-\frac{1}{4\pi}\int_0^{2\pi} g_1(\rho e^{i\theta})\Big\{\frac{\cos^2(\theta - \varphi)}{a\sin(\theta - \varphi)}\operatorname{arctg}\frac{a\sin(\theta - \varphi)}{1 - a\cos(\theta - \varphi)} \\ &- \frac{1}{2} - \frac{\cos(\theta - \varphi)}{2a}\ln|1 - 2a\cos(\theta - \varphi) + a^2|\Big\}\, d\theta. \end{aligned} \qquad (3.31)$$

We recall that $a = \rho/r$, hence

$$\begin{aligned} v(\rho e^{i\varphi}) = &-\frac{1}{4\pi}\int_0^{2\pi} g_1(\rho e^{i\theta})\Big\{\frac{r\cos^2(\theta - \varphi)}{\rho\sin(\theta - \varphi)}\operatorname{arctg}\frac{\rho\sin(\theta - \varphi)}{r - \rho\cos(\theta - \varphi)} \\ &- \frac{1}{2} - \frac{r\cos(\theta - \varphi)}{2\rho}\ln\frac{r^2 - 2r\rho\cos(\theta - \varphi) + \rho^2}{r^2}\Big\}\, d\theta. \end{aligned} \qquad (3.32)$$

From (3.15), (3.16) and from the equalities

$$\rho = |x - x_0|, \quad |x - y|^2 = r^2 + \rho^2 - 2r\rho\cos(\theta - \varphi),$$

we get the desired relation (3.13).

The theorem is proved. □

Corollary 3.1.

At the central point of the circle $K(x_0, r)$, the biharmonic function satisfies the spherical mean value relation

$$\begin{aligned} u(x_0) &= \frac{1}{r\omega_2}\int_{S(x_0,r)} u(y)dS_y - \frac{r}{4\omega_2}\int_{S(x_0,r)} \Delta u(y)dS_y\,, \\ \Delta u(x_0) &= \frac{1}{r\omega_2}\int_{S(x_0,r)} \Delta u(y)dS_y\,. \end{aligned} \qquad (3.33)$$

Note that this relation was obtained above (see the formula (3.12)) through series expansion.

3.2.3. Rigid fixing of the boundary

The generalized Poisson formula can be obtained also for biharmonic functions with rigid fixing on the boundary, i.e., for the solutions to the boundary value problem (3.9),(3.8).

Theorem 3.2.

The biharmonic function satisfies at arbitrary point $x \in K(x_0, r) \subset G$ the following spherical mean value relation:

$$\begin{aligned} u(x) &= -\frac{r^2 - |x - x_0|^2}{2r^2\omega_2} \int\limits_{S(x_0,r)} \frac{g_1(y)dS_y}{|x-y|^2} \\ &+ \frac{(r^2 - |x-x_0|^2)^2}{2r^3\omega_2} \int\limits_{S(x_0,r)} \frac{(2r^2 - 2r|x-x_0|\cos\alpha)g_0(y)dS_y}{|x-y|^4} , \end{aligned} \qquad (3.34)$$

where α is the angle between the vectors $x - x_0$ and $y - x_0$.

Proof. Again, we use the representation (3.15). First, from the boundary condition we have

$$w(y) = u(y) = g_0(y), \quad y \in S(x_0, r).$$

The Poisson formula yields

$$w(x) = \frac{r^2 - t^2}{r\omega_2} \int \frac{g_0(y)dS_y}{\rho^2} , \qquad (3.35)$$

where $\rho = |x - y|$, $t = |x - x_0|$. From (3.15) it follows that

$$\frac{\partial u}{\partial t} = 2tv + (t^2 - r^2)\frac{\partial v}{\partial t} + \frac{\partial w}{\partial t} ,$$

and $\partial/\partial t = -\partial/\partial n$ as $y \to y' \in S(x_0, r)$.

Hence,

$$-2rv(y) + \frac{\partial w}{\partial n}(y) = g_1(y), \quad y \in S(x_0, r). \qquad (3.36)$$

Let

$$h(x) = v(x) + \frac{t}{2r^2}\frac{\partial w}{\partial t} \, .$$

From (3.36) it follows that for $y \in S(x_0, r)$

$$h(y) = v(y) - \frac{1}{2r}\frac{\partial w}{\partial n} = -\frac{g_1(y)}{2r} \, ,$$

and hence for the function $h(x)$ we can write

$$h(x) = -\frac{r^2 - t^2}{2r^2\omega_2} \int\limits_{S(x_0,r)} \frac{g_1(y)dS_y}{\rho^2} \, .$$

On the other hand,

$$\frac{\partial w}{\partial t} = -\frac{2t}{r\omega_2} \int\limits_{S(x_0,r)} \frac{g_0(y)dS_y}{\rho^2} - \frac{2(r^2 - t^2)}{r\omega_2} \int\limits_{S(x_0,r)} \frac{(t - r\cos\alpha)g_0(y)dS_y}{\rho^4}$$

which proves the theorem. □

Corollary 3.2.

At the center of $S(x_0, r)$, the mean value relation (3.34) takes the form

$$u(x_0) = -\frac{1}{2\omega_2} \int\limits_{S(x_0,r)} g_1(y)dS_y + \frac{1}{r\omega_2} \int\limits_{S(x_0,r)} g_0(y)dS_y \, . \tag{3.37}$$

Note that the mean value relation (3.35) can be differentiated with respect to x. Let $x = (x_1, x_2)$, $x_0 = (x_1^0, x_2^0)$, $y = (y_1, y_2)$. Then

$$\begin{aligned}
\Delta u(x) = {} & \frac{4}{r\omega_2}(r^2 - 2t^2) \int\limits_{S(x_0,r)} \frac{g_1(y)dS_y}{\rho^2} - \frac{2(r^2 - t^2)^2}{r^2\omega_2} \int\limits_{S(x_0,r)} \frac{g_1(y)dS_y}{\rho^4} \\
& + k(x_1, x_1^0) \int\limits_{S(x_0,r)} A(x_1, x_1^0, y_1)g_1(y)dS_y \\
& + k(x_2, x_2^0) \int\limits_{S(x_0,r)} A(x_1, x_2^0, y_2)g_1(y)dS_y \\
& - \frac{8}{r^2\omega_2}(r^2 - 2t^2) \int\limits_{S(x_0,r)} \frac{r - t\cos\alpha}{r^4} g_0(y)dS_y
\end{aligned}$$

$$+\frac{(r^2-t^2)^2}{r^3\omega_2}\int\limits_{S(x_0,r)} B(y)g_0(y)dS_y$$

$$-\frac{k(x_1,x_1^0)}{r}\int\limits_{S(x_0,r)} A(x_1,x_1^0,y_1)\ g_0(y)dS_y$$

$$-\frac{k(x_2,x_2^0)}{r}\int\limits_{S(x_0,r)} A(x_2,x_1^0,y_2)\ g_0(y)dS_y\ ,$$

where $t=|x-x_0|,\ \rho=|x-y|$,

$$k(z_1,z_2)=\frac{2}{r\omega_2}(r^2-t^2)(z_1-z_2),$$

$$A(p,q,z)=(2(q-p)\ r^2-8(r^2-rt\cos\alpha)(p-z))/\rho^6,$$

and

$$B(y)=[16r(r-t\cos\alpha)-8r^2+5((x_1-y_1)(x_1-x_1^0)$$

$$+(x_2-y_2)(x_2-x_2^0))]/\rho^6.$$

Corollary 3.3.

At the central point,

$$\Delta u(x_0)=\frac{2}{r^2\omega_2}\int\limits_{S(x_0,r)} g_1(y)dS_y\ .$$

Let us consider now the boundary value problem

$$\Delta\Delta u(x)=0,\quad x\in K(x_0,r)\subset R^2, \tag{3.38}$$

$$u(y)=g_0(y),\ \Delta u(y)+\frac{1-\nu}{\rho}\frac{\partial u(y)}{\partial n}=g_1(y),\quad y\in S(x_0,r). \tag{3.39}$$

From the above two theorems the following mean value relation for the problem (3.38), (3.39) is true:

$$u(x_0) = \frac{1}{r\omega_2} \int\limits_{S(x_0,r)} g_0(y)dS_y - \frac{r}{4\omega_2}\left(1+\frac{\alpha r}{2}\right)^{-1} \int\limits_{S(x_0,r)} g_1(y)dS_y \,,$$

$$\Delta u(x_0) = \frac{1}{r\omega_2}\left(1+\frac{\alpha r}{2}\right)^{-1} \int\limits_{S(x_0,r)} g_1(y)dS_y \,, \tag{3.40}$$

where $\alpha = (1-\nu)/\rho$.

3.2.4. Nonhomogeneous biharmonic equation

Let us consider a nonhomogeneous biharmonic equation,

$$\Delta^2 u = g.$$

For this equation, we now derive mean value relations in the circle $K(x,r)$. To do this we use the equality [40]

$$N_r u = \sum_{j=0}^{m} \left(\frac{r}{2}\right)^{2j} \frac{\Delta^j u(x)}{(j!)^2} + B_x^r(v_m \Delta^{m+1} u), \tag{3.41}$$

where $B_x^r(u)$ is the averaging operator of $u(x)$ over the circle $K(x,r)$, i.e.,

$$B_x^r(u) = \int\limits_{K(x,r)} u d\sigma_x^r.$$

The functions v_m are determined from the recurrence formulae

$$v_0(\rho) = \frac{1}{2\pi} \ln\frac{r}{\rho} \,, \quad v_{j+1} = \int\limits_0^r \rho v_j(\rho) \ln\frac{\rho}{r} \, d\rho \,. \tag{3.42}$$

Hence using $\Delta^2 u = g$ we obtain from (3.41) for $m = 0$

$$\Delta u(x) = N_r(\Delta u) + B_x^r\left(-\frac{g}{2\pi}\ln\frac{r}{r_1}\right).$$

Integrating (3.42) by parts we find

$$v_1 = \int_{r_1}^{r} \rho v_0(\rho) \ln \frac{\rho}{\rho_1} d\rho$$
$$= \frac{1}{2\pi}\left(\frac{r^2 + r_1^2}{4} \ln \frac{r}{r_1} - \frac{r^2 - r_1^2}{4}\right).$$

From this we get, using (3.41) for $m = 1$, the identity

$$u(x) = N_r(u) - \frac{r^2}{4} N_r(\Delta u)$$
$$+ B_x^r\left(\frac{1}{2\pi}\left\{\frac{r^2 - r_1^2}{4} - \frac{r_1^2}{4} \ln \frac{r}{r_1}\right\} g\right) .$$

Hence, the formulae (3.43),(3.43) present a generalized mean value relation in the case when the right-hand side is not equal to zero.

For plates with rigid fixing of the boundary, an analogous mean value relation can be derived.

Theorem 3.3.

An arbitrary regular solution to the equation $\Delta u = g$ satisfies the mean value relation

$$u(x) = N_r(u) - \frac{r}{2} N_r\left(\frac{\partial u}{\partial r}\right) + B_x^r(q_1 g) , \tag{3.44}$$

$$\Delta u(x) = \frac{2}{r} N_r\left(\frac{\partial u}{\partial r}\right) + B_x^r(q_2 g), \tag{3.45}$$

where

$$q_1(r_1) = \frac{1}{2\pi}\left(\frac{r^2 + r_1^2}{8} - \frac{r_1^2}{4} \ln\left(\frac{r}{r_1}\right)\right),$$

$$q_2(r_1) = -\frac{1}{2\pi}\left(\ln\left(\frac{r}{r_1}\right) - \frac{1}{2} + \frac{r_1^2}{2r^2}\right),$$

and the integration B_x^r of the functions $q_1(r_1)g(r_1)$ and $q_2(r_1)g(r_1)$ is carried out over the circle $K(x,r)$.

Proof. By the second Green formula

$$B_x^r(\Delta u) = 2\pi r \cdot N_r\left(\frac{\partial u}{\partial r}\right).$$

Substituting this expression into the Poisson formula for $\Delta u(x)$,

$$\Delta u(x) = \frac{1}{\pi r^2} B_x^r(\Delta u) - \frac{1}{2\pi} B_x^r \left(\left[r^2 \ln(r/r_1) - \frac{r^2 - r_1^2}{2} \right] g \right), \tag{3.46}$$

we obtain (3.45). Substituting now (3.46) into (3.41) for $m = 1$ we obtain by (3.45)

$$\begin{aligned} u(x) &= N_r(u) - \frac{r^2}{4} \Delta u(x) - B_x^r(v_1 g) \\ &= N_r(u) - \frac{r}{2} N_r\left(\frac{\partial u}{\partial r}\right) + \frac{1}{8\pi} B_x^r \left(\left[r^2 \ln(r/r_1) - \frac{r^2 - r_1^2}{2} \right] g \right) \\ &- B_x^r(v_1 g) \\ &= N_r(u) - \frac{r}{2} N_r\left(\frac{\partial u}{\partial r}\right) + B_x^r(q_1 g). \end{aligned}$$

This proves (3.44). □

Remark 3.1.

Note that from the two mean value relations obtained above we can deduce a mean value relation for the case of simply supported boundary of the plate. Indeed, combining these relations we find that an arbitrary regular solution to $\Delta^2 u = g$ satisfies the mean value relation

$$u(x) = N_r(u) - \frac{r^2}{4 - 2\alpha r} N_r\left(\Delta u - \alpha \frac{\partial u}{\partial r}\right) + \frac{2}{2 - \alpha r} B_x^r(q_1' g),$$

and

$$\Delta u(x) = \frac{2}{2 - \alpha r} N_r\left(\Delta u - \alpha \frac{\partial u}{\partial r}\right) + \frac{2}{2 - \alpha r} B_x^r(q_2' g),$$

where α is an arbitrary constant, and

$$\begin{aligned} q_1' &= \frac{1}{2\pi} \left(\frac{r^2 - r_1^2}{4} - \frac{r^2}{4} \ln(r/r_1) \right) - \frac{\alpha r}{2} q_1, \\ q_2' &= -\frac{1}{2\pi} \ln(r/r_1) \frac{\alpha r}{2} q_2. \end{aligned}$$

3.3. Fourth order equation governing the bending of a plate on an elastic base surface

Let us consider the following fourth order equation with constant coefficients:

$$\Delta\Delta u(x) + a_1 \Delta u(x) + a_2 u(x) = 0, \quad x \in K(x_0, r), \tag{3.47}$$

with the boundary conditions

$$u(y) = g_0(y), \quad \frac{\partial u}{\partial n}(y) = g_1(y), \quad y \in S(x_0, r). \tag{3.48}$$

In (3.47) a_1, a_2 are constants; hence it can be rewritten in the form

$$(\Delta - k_1)(\Delta - k_2)u(x) = 0 \ .$$

In what follows it is assumed that $k_1 \neq k_2$ and $\operatorname{Re} k_i > \lambda_0$, $i = 1, 2$, where λ_0 is the principal eigenvalue of the Laplace operator. The case $k_1 = k_2$ is treated in Section 3.4.1.

In [14], the spherical mean value relation was obtained for the boundary conditions

$$u(y) = g_0(y), \quad \Delta u(y) = g_1(y). \tag{3.49}$$

Theorem 3.4.

The solution to (3.47) satisfies the following spherical mean value relation

$$u(x_0) = \frac{c_{11}(r)}{2\pi} \int_0^{2\pi} u(x_0 + re^{i\theta})d\theta + \frac{c_{12}(r)}{2\pi} \int_0^{2\pi} \Delta u(x_0 + re^{i\theta})d\theta, \tag{3.50}$$

$$\Delta u(x_0) = \frac{c_{21}(r)}{2\pi} \int_0^{2\pi} u(x_0 + re^{i\theta})d\theta + \frac{c_{22}(r)}{2\pi} \int_0^{2\pi} \Delta u(x_0 + re^{i\theta})d\theta, \tag{3.51}$$

where

$$c_{11}(r) = d_{11}(I_0)/\delta, \quad c_{12}(r) = d_{12}(I_0)/\delta,$$

$$c_{21}(r) = d_{21}(I_0)/\delta, \quad c_{22}(r) = d_{22}(I_0)/\delta,$$

and

$$\begin{aligned}
\delta &= (k_1^2 - k_2^2) I_0(k_1 r) I_0(k_2 r), \\
d_{11}(I_n) &= k_1^2 I_n(k_1 r) - k_2^2 I_n(k_2 r), \\
d_{12}(I_n) &= -(I_n(k_1 r) - I_n(k_2 r)), \\
d_{21}(I_n) &= k_1^2 k_2^2 (I_n(k_1 r) - k_1^2 I_n(k_2 r)), \\
d_{22}(I_n) &= -(k_2^2 I_n(k_1 r) - k_1^2 I_n(k_2 r)),
\end{aligned}$$

and I_n is the modified Bessel function ($I_n(z) = J_n(iz)i^{-n}$), where J_z is the standard Bessel function (see the power expansion (2.6) *of this function in Section 2.1.1).*

Theorem 3.5.

The solution (in $K(x_0, r) \subset \mathbb{R}^2$) to (3.47) *satisfies the mean value relation*

$$u(x_0) = \frac{b_{11}(r)}{r\omega_2} \int\limits_{S(x_0,r)} u(y)dS_y + \frac{b_{12}(r)}{r\omega_2} \int\limits_{S(x_0,r)} \frac{\partial u}{\partial n}(y)dS_y, \tag{3.52}$$

$$\Delta u(x_0) = \frac{b_{21}(r)}{r\omega_2} \int\limits_{S(x_0,r)} u(y)dS_y + \frac{b_{22}(r)}{r\omega_2} \int\limits_{S(x_0,r)} \frac{\partial u}{\partial n}(y)dS_y, \tag{3.53}$$

where

$$\begin{aligned} &b_{11}(r) = (d_{11}(I_0) - d_{11}(I_2))/\delta_1, \quad b_{12}(r) = d_{12}(I_0)/\delta_1, \\ &b_{21}(r) = (d_{21}(I_0) - d_{21}(I_2))/\delta_1, \quad b_{22}(r) = d_{22}(I_0)/\delta_1, \\ &\delta_1 = \delta + S, \quad S = k_1^2 I_0(k_2 r) I_2(k_1 r) - k_2^2 I_0(k_1 r) I_2(k_2 r), \end{aligned} \tag{3.54}$$

□

Proof. The general solution to (3.47) at a point x with polar coordinates (r, θ) can be written as

$$\begin{aligned} u(r,\theta) = A_0 I_0(k_1 r) + B_0 I_0(k_2 r) + \sum_{n=1}^{\infty} (A_n I_n(k_1 r) + B_n I_n(k_2 r)) \sin n\theta \\ + \sum_{n=1}^{\infty} (C_n I_n(k_1 r) + D_n I_n(k_2 r)) \cos n\theta, \end{aligned} \tag{3.55}$$

where A_n, B_n, C_n, D_n ($n \geq 0$) are arbitrary constants. We differentiate this representation with respect to r (which corresponds to the differentiation along the radius of the circle) and use the known formulae

$$2I_n' = I_{n-1} + I_{n+1}, \quad I_0' = I_1.$$

This leads to

$$\frac{\partial u}{\partial r}(r,\theta) = A_0 k_1 I_1(k_1 r) + B_0 k_2 I_1(k_2 r)$$
$$+\sum_{n=1}^{\infty}[A_n \frac{k_1}{2}(I_{n-1}(k_1 r) + I_{n+1}(k_1 r))$$
$$+ B_n \frac{k_2}{2}(I_{n-1}(k_2 r) + I_{n+1}(k_2 r))] \sin n\theta \qquad (3.56)$$
$$+\sum_{n=1}^{\infty}[C_n \frac{k_1}{2}(I_{n-1}(k_1 r) + I_{n+1}(k_1 r))$$
$$+ D_n \frac{k_2}{2}(I_{n-1}(k_2 r) + I_{n+1}(k_2 r))] \cos n\theta \ .$$

Integrating (3.55) and (3.56) over $(0 \le \theta \le 2\pi)$ yields

$$\frac{1}{2\pi}\int_0^{2\pi} u(r,\theta)d\theta = A_0 I_0(k_1 r) + B_0 I_0(k_2 r), \qquad (3.57)$$

$$\frac{1}{2\pi}\int_0^{2\pi} \frac{\partial u}{\partial r}(r,\theta)d\theta = A_0 k_1 I_1(k_1 r) + B_0 k_2 I_1(k_2 r) \ . \qquad (3.58)$$

We multiply the last equality by $2/r$ and use the known relation $2I_n = x(I_{n-1} - I_{n+1})$ for $n = 1$ (e.g. see [7]). This gives

$$\Delta u(x_0) = A_0 k_1^2 (I_0(k_1 r) - I_2(k_1 r)) + B_0 k_2^2 (I_0(k_2 r) - I_2(k_2 r)) \ . \qquad (3.59)$$

We used here also the relation

$$\Delta u(x_0) = \frac{1}{2\pi}\frac{2}{r}\int_0^{2\pi} \frac{\partial u}{\partial r}(r,\theta)d\theta \ .$$

Let us consider the relations (3.57) (3.59) at the point x_0, the center of the circle. Note that $I_0(0) = 1$ and $I_2(0) = 0$. Therefore,

$$u(x_0) = A_0 + B_0, \qquad (3.60)$$

$$\Delta u(x_0) = k_1^2 A_0 + k_2^2 B_0 \ . \qquad (3.61)$$

Let

$$S_1 = \frac{1}{2\pi}\int\limits_0^{2\pi} u(r,\theta)d\theta, \quad S_2 = \frac{1}{2\pi}\frac{2}{r}\int\limits_0^{2\pi} \frac{\partial u}{\partial r}(r,\theta)d\theta.$$

We resolve (3.57) (3.59) with respect to the unknown constant A_0

$$\begin{aligned} A_0 &= \frac{1}{\delta_1}\begin{vmatrix} S_1 & I_0(k_2 r) \\ S_2 & k_2^2(I_0(k_2 r) - I_2(k_2 r)) \end{vmatrix} \\ &= (S_1 k_2^2(I_0(k_2 r) - I_2(k_2 r)) - S_2 I_0(k_2 r))/\delta_1, \end{aligned}$$

and B_0

$$\begin{aligned} B_0 &= \frac{1}{\delta_1}\begin{vmatrix} I_0(k_1 r) & S_1 \\ k_1^2(I_0(k_1 r) - I_2(k_1 r)) & S_2 \end{vmatrix} \\ &= (S_2 I_0(k_1 r) - k_1^2 S_1(I_0(k_1 r) - I_2(k_1 r)))/\delta_1\ , \end{aligned}$$

where

$$\begin{aligned} \delta_1 &= \begin{vmatrix} I_0(k_1 r) & I_0(k_2 r) \\ k_1^2(I_0(k_1 r) - I_2(k_1 r)) & k_2^2(I_0(k_2 r) - I_2(k_2 r)) \end{vmatrix} \\ &= \delta - k_2^2 I_0(k_1 r) I_2(k_2 r) + I_0(k_2 r) I_2(k_1 r) = \delta + S. \end{aligned}$$

□

Corollary 3.4.

The solution to (3.47) satisfies the mean value relation (in the circle $K(x_0, r) \subset \mathbb{R}^2$)

$$u(x_0) = \frac{e_{11}(r)}{r\omega_2}\int\limits_{S(x_0,r)} u(y)dS_y + \frac{e_{12}(r)}{r\omega_2}\int\limits_{S(x_0,r)} (\Delta u(y) + \alpha\frac{\partial u}{\partial n}(y))dS_y, \quad (3.62)$$

$$\Delta u(x_0) = \frac{e_{21}(r)}{r\omega_2}\int\limits_{S(x_0,r)} u(y)dS_y + \frac{e_{22}(r)}{r\omega_2}\int\limits_{S(x_0,r)} (\Delta u(y) + \alpha\frac{\partial u}{\partial n}(y))dS_y, \tag{3.63}$$

where

$$\begin{aligned} e_{11}(r) &= ((1+\frac{\alpha r}{2})d_{11}(I_0) + \frac{\alpha r}{2}d_{11}(I_2))/\delta_2, \\ e_{12}(r) &= d_{12}(I_0)/\delta_2, \\ e_{21}(r) &= ((1+\frac{\alpha r}{2}d_{21})(I_0) + \frac{\alpha r}{2}d_{21}(I_2))/\delta_2, \\ e_{22}(r) &= d_{22}(I_0)/\delta_2, \\ \delta_2 &= (1+\frac{\alpha r}{2})\delta + \frac{\alpha r}{2}S, \end{aligned} \tag{3.64}$$

and the coefficients δ, s, d_{ij} are defined in (3.54), *α is an arbitrary constant.*

3.4. Metaharmonic equations

We introduce the metaharmonic operator in $\mathbb{R}^m$,

$$L(\Delta) = (\Delta - \lambda_n)^{p_n}(\Delta - \lambda_{n-1})^{p_{n-1}} \dots (\Delta - \lambda_1)^{p_1},$$

where $p_1 \le p_2 \le \dots \le p_n$ are positive integers, $\lambda_1, \lambda_2, \dots, \lambda_n$ are real or complex numbers; it is assumed that $\operatorname{Re}\lambda_i$, the real parts of the constants λ_i, $i = 1, 2, \dots, n$ are larger than the principal eigenvalue of the Laplace operator. The integer number $\nu = p_1 + p_2 + \dots + p_n$ is the order of the metaharmonic operator.

3.4.1. Polyharmonic equation

We first derive a spherical mean value relation for the polyharmonic equation

$$\Delta^p v(x) = 0$$

in $\mathbb{R}^m$.

To this end we apply the representation (3.10). Since $\Delta^p u = 0$, this representation reads

$$v(x_0) = -\Gamma(\frac{m}{2}) \sum_{i=0}^{p-1} \left(\frac{r}{2}\right)^{2i} \frac{\Delta^i v(x_0)}{i!\Gamma(i+m/2)} + N_r v(x),$$

where x_0 is the center of the sphere $S(x_0, r)$ in $\mathbb{R}^m$ and N_r is the spherical mean over this sphere. Successive use of this representation yields the following triangular system

$$\Delta^{p-1}\, v\, (x_0) = N_r \Delta^{p-1} v,$$
$$\Delta^{p-2}\, v\, (x_0) = -c_1 \Delta^{p-1} v(x_0) + N_r \Delta^{p-2} v,$$
$$\vdots$$

etc., the last equation coinciding with (3.10):

$$v(x_0) = -c_1 \Delta v(x_0) - c_2 \Delta^2 v(x_0) - \ldots - c_{p-1} \Delta^{p-1} v(x_0) + N_r v.$$

Here

$$c_i = \frac{\Gamma(m/2)(r/2)^{2i}}{i!\Gamma(i+m/2)}.$$

We rewrite this triangular system in the form

$$\Delta^j v(x_0) = -\sum_{i=j+1}^{p-1} \Delta^i v(x_0) + N_r \Delta^j v, \quad j = 0, 1, \ldots, p-1,$$

where $\Delta^0 v \equiv v$.

Substituting in this system the first equation into the second one, then the second into the third one, etc., we obtain

$$\Delta^{p-1}\, v\, (x_0) = N_r \Delta^{p-1}\, v,$$
$$\Delta^{p-2}\, v\, (x_0) = N_r \Delta^{p-2}\, v - s_1 N_r \Delta^{p-1}\, v,$$
$$\vdots$$

etc., with the last equation

$$v(x_0) = N_r v - s_1 N_r \Delta v - \ldots - s_{p-1} N_r \Delta^{p-1} v.$$

Here the coefficients s_i are obtained by the recurrent formula:

$$s_1 = c_1, \; s_i = c_i - \sum_{j=1}^{i-1} c_j s_{i-j}, \quad i > 1.$$

Thus we obtained the desired spherical mean value relation

$$\Delta^j v(x_0) = N_r \Delta^j v - \sum_{i=j+1}^{p-1} s_i N_r \Delta^i v, \quad j = 0, 1, \ldots, p-1 \tag{3.65}$$

which can be written in a matrix form

$$\mathbf{v} = N_r[K(r)\mathbf{v}] \; ,$$

where $\mathbf{v} = (v, \Delta v, \ldots, \Delta^{p-1} v)$, and $K(r)$ is the triangular matrix

$$K(r) = \begin{pmatrix} 1 & -s_1 & -s_2 & \cdots & -s_{p-1} \\ 0 & 1 & -s_1 & \cdots & -s_{p-2} \\ \cdots & \cdots & \cdots & \cdots & \cdots \\ \cdots & \cdots & \cdots & 1 & -s_1 \\ \cdots & \cdots & \cdots & 0 & 1 \end{pmatrix} .$$

Let us consider the following boundary value problem for the polyharmonic equation in a bounded domain G:

$$\begin{aligned} &\Delta^p v(x) = 0, \quad x \in G, \quad v \in C^{2p}(G) \cap C^{2p-2}(\bar{G}), \\ &\mathbf{v}|_\Gamma(y) = \mathbf{g}(y), \quad y \in \Gamma, \end{aligned} \tag{3.66}$$

where

$$\mathbf{v} = (v, \Delta v, \ldots, \Delta^{p-1} v), \; \mathbf{g} = (g_1, \ldots, g_p) = (v, \Delta v, \ldots, \Delta^{p-1} v)|_\Gamma.$$

Theorem 3.6.

Assume that the boundary value problem (3.66) *is uniquely solvable for each continuous* $\mathbf{g}$. *Then if a function v from the class $C^{2p}(G) \cap C^{2p-2}(\bar{G})$ satisfies the spherical mean value relation* (3.65) *at all points $x \in G$ for at least one sphere $S(x, r) \subset G$ then u is the unique solution to* (3.66).

The proof repeats the arguments given in Theorem 2.5 and Proposition 3.1.

Remark 3.2.

As mentioned in the Introduction, it is possible to derive different spherical mean value relations for one and the same equation. For instance, the polyharmonic functions ($\Delta^p v = 0$) satisfy also the following spherical mean value relation at an arbitrary point $x \in B(x_0, r)$ [50]:

$$v(x) = P(x,r) \sum_{j=1}^{p} \int_{S(x_0,r)} \binom{p-1}{j-1} g_j(y) \left[\frac{\partial^{p-j}}{\partial(r^2)^{p-j}} \frac{r^{m-2}}{|x-y|^m} \right] do(y),$$

where

$$P(x,r) = \frac{(-1)^{p-1}(r^2 - |x-x_0|^2)^p}{(p-1) r^{m-1} \omega_m}$$

and

$$g_j = \left. \frac{\partial^{j-1} v}{\partial n^{j-1}} \right|_{\Gamma}, \quad j = 1, \dots, p.$$

It is however not clear if this relation characterizes the relevant boundary value problem for the polyharmonic equation. The same is true for another mean value relation for the polyharmonic equation obtained in [30]

$$v(x) = \sum_{q=0}^{p-1} c(m,p,q) \kappa_{q,x}^r(v),$$

where the means $\kappa_{q,x}^r(v)$ are obtained recursively starting from the uniform measure $\sigma_x^r(v)$ over the sphere $S(x,r)$:

$$\kappa_{q,x}^r(v) = \frac{m}{r^m} \int_0^r t^{m-1} \kappa_{q-1,x}^r(v)\, dt, \quad q = 1, 2 \dots,$$

and the constants $c(m,p,q)$ are known (e.g. see [30]).

3.4.2. General case

Let us consider now the general case of the metaharmonic equation. We start with the equation $(\Delta - \lambda)^p u = 0$, such that the constant λ (real or complex) satisfies the condition: $\operatorname{Re} \lambda > \lambda_0$ where λ_0 is the principal eigenvalue of the Laplace equation. Next we show that the general case of the metaharmonic equation is derived from this particular case.

Theorem 3.7.

The solution of the equation

$$(\Delta - \lambda)^p u = 0$$

satisfies the mean value relation

$$v(x_0) = N_r(Av). \tag{3.67}$$

Here

$$v = (u, (\Delta - \lambda)u, \dots, (\Delta - \lambda)^{p-1}u)^T,$$
$$A = (I + \beta_1 H + \dots + \beta_{p-1} H^{p-1}) w_m^{-1}(r, \lambda),$$

where the entries $\beta_1, \dots, \beta_{p-1}$ are determined from the recursive relation

$$\beta_1 = -\alpha_1, \beta_2 = -\alpha_2 - \beta_1\alpha_1, \dots, \beta_k = -\alpha_k - \alpha_{k-1}\beta_1 - \dots - \alpha_1\beta_{k-1}, \tag{3.68}$$

I is a $p \times p$ identity matrix, H is a matrix whose non-zero entries have the form $h_{i,i+1} = 1$.

In (3.68)

$$\alpha_i = \gamma_i / w_m(r, \lambda), \quad i = 1, \dots, p-1,$$

where

$$\gamma_k = \frac{1}{k!}\frac{\partial^k w_m}{\partial \lambda^k}(r, \lambda) = \frac{r^{2k}}{k!}\frac{w_{m+2k}(r, \lambda)}{2^k m(m+2)\dots(m+2k-2)}.$$

Proof. Expanding in (2.7) the function $W_m(ir\lambda^{1/2})$ in a power series in $(\Delta - \lambda)$, we obtain, using the relation $(\Delta - \lambda)^p u = 0$, (e.g. see [34])

$$N_r(u) = \sum_{k=1}^{p} \frac{1}{(k-1)!}\frac{\partial^{k-1} w_m(r, \lambda)}{\partial \lambda^{k-1}}(\Delta - \lambda)^{k-1} u(x_0). \tag{3.69}$$

Using relations similar to (3.69) for the functions $(\Delta - \lambda)u, (\Delta - \lambda)^2 u, \dots,$ and $(\Delta - \lambda)^{p-1}u$, we obtain a system of equations which can be written in a matrix form,

$$N_{x_0}^r(v) = Rv(x_0), \tag{3.70}$$

where $R = \{r_{ij}\}_{i,j=1}^p$ is a matrix with diagonal entries

$$\gamma_0 = r_{ii} = w_m(r,\lambda), \quad i = 1,2,\dots,p$$

and non-zero entries

$$\gamma_1 = r_{i,i+1} = \frac{1}{1!}\frac{\partial w_m}{\partial\lambda}(r,\lambda) = \frac{1}{1!}\frac{r^2}{2m}w_{m+2}(r,\lambda) ,$$

$$\gamma_2 = r_{i,i+2} = \frac{1}{2!}\frac{\partial^2 w_m}{\partial\lambda^2}(r,\lambda) = \frac{1}{2!}\frac{r^4 w_{m+4}(r,\lambda)}{2^2 m(m+2)} ,$$

$$\vdots$$

$$\vdots$$

etc.,

$$\gamma_k = r_{i,i+k} = \frac{1}{k!}\frac{\partial^k w_m}{\partial\lambda^k}(r,\lambda) = \frac{1}{k!}\frac{r^{2k}w_{m+2k}(r,\lambda)}{2^k m(m+2)\dots(m+2k-2)},$$

where $i,k = 1,2,\dots,p-1$.

Consequently, using the specific structure of R, this matrix can be written in the form

$$R = (I + \alpha_1 H + \alpha_2 H^2 + \dots + \alpha_{p-1}H^{p-1})w_m,$$
$$\alpha_i = \frac{\gamma_i}{w_m}, \quad i = 1,\dots,p-1,$$

where the matrix H is defined in the theorem. It is clear that the matrix $A = R^{-1}$ has the same structure, i.e.,

$$R^{-1} = (I + \beta_1 H + \beta_2 H^2 + \dots + \beta_{p-1}H^{p-1})w_m^{-1}.$$

From $R^{-1}R = I$ we obtain (using the condition $H^k = 0$ for $k \geq p$) the recurrent relations (3.68) for the entries $\beta_1,\dots,\beta_{p-1}$. Hence, the matrix $A = R^{-1}$ is easily constructed explicitly, and we have from (3.70)

$$v(x_0) = N_r(Av)$$

which coincides with (3.67).

Consider the following boundary value problem for the metaharmonic equation in a domain $G \subset \mathbb{R}^m$

$$Lu(x) = 0, \quad x \in G\ , \tag{3.71}$$

$$u|_{\partial G} = g_1, \ldots, \Delta u|_{\partial G} = g_2, \ldots, \Delta^{\nu-1} u|_\Gamma = g_\nu\ , \tag{3.72}$$

where ν is the order of the metaharmonic operator L. Here and in the following discussion $\mathbf{v}|_\Gamma = \mathbf{g}$ means that

$$\mathbf{v}(x) \to \mathbf{g}, \text{ as } x \to \xi \in \Gamma, \quad (x \in G).$$

It is not difficult to pass from (3.71), (3.72) to the equation

$$Lu(x) = 0, \quad x \in G,$$

with the boundary conditions

$$u|_{\partial G} = \psi_1, \quad (\Delta - \lambda_1) u|_\Gamma = \psi_2, \ldots, (\Delta - \lambda_1)^{p_1 - 1} u|_\Gamma = \psi_{p_1}, \ldots,$$
$$(\Delta - \lambda_n)^{p_n - 1} (\Delta - \lambda_{n-1})^{p_{n-1}} \ldots (\Delta - \lambda_1)^{p_1} u|_\Gamma = \psi_\nu\ ,$$

where ψ_i are linearly dependent on $g_1, g_2, \ldots, g_\nu$.

We will seek the solution to (3.71),(3.72) in the form

$$u = u_1 + u_2 + \ldots + u_n \tag{3.73}$$

where u_i satisfies the equation

$$(\Delta - \lambda_i)^{p_i} u = 0, \quad i = 1, 2, \ldots, n. \tag{3.74}$$

From (3.71),(3.72) it is obvious that $Lu = 0$. To obtain a unique representation of the solution $u(x)$ in the form of (3.73), it is necessary to derive the appropriate boundary conditions for (3.74).

We introduce the notation

$$\begin{aligned} &(\Delta - \lambda_i)^{l-1} u_i = v_i^{(l)}\ , \\ &v_i^{(l)}|_\Gamma = \psi_{il}, \quad i = 1, 2, \ldots, n, \quad l = 1, 2, \ldots, p_i \end{aligned} \tag{3.75}$$

and consider vectors of dimension ν:

$$V = \left(v_1^{(1)}, v_1^{(2)}, \ldots, v_1^{(p_1)}, v_2^{(1)}, v_2^{(2)}, \ldots, v_2^{(p_2)}, \ldots, v_n^{(1)}, v_n^{(2)}, \ldots, v_n^{(p_n)}\right)\ ,$$
$$\psi = \left(\psi_{11}, \psi_{12}, \ldots, \psi_{1p_1}, \psi_{21}, \psi_{22}, \ldots, \psi_{2p_2}, \ldots, \psi_{n1}, \psi_{n2}, \ldots, \psi_{np_n}\right)\ ,$$
$$\Psi = (\Psi_1, \ldots, \Psi_\nu)\ .$$

Let

$$\mu_1 = \lambda_1,\ \mu_2 = \lambda_1, \ldots, \mu_{p_1} = \lambda_1;\ \mu_{p_1+1} = \lambda_2, \ldots, \mu_{p_1+p_2} = \lambda_2; \ldots$$

$$\mu_{p_1+p_2+\ldots,+p_{n-1}+1} = \lambda_n, \ldots, \mu_\nu = \lambda_n$$

and define a vector $U = (U_1, \ldots, U_\nu)$ with $U_1 = u$ and

$$U_j = \prod_{i=1}^{j-1} (\Delta - \mu_i) u,\ j = 2, \ldots, \nu.$$

We derive a linear relation between the vector V and U in the form

$$BV = U, \tag{3.76}$$

i.e., the problem is to find the matrix B. To do this, we apply to (3.74) the operator $\prod_{i=1}^{k}(\Delta - \mu_i)$ for $k = 0, 1, \ldots, \nu - 1$ ($k = 0$ corresponds to the identity operator). Hence,

$$U_1 = v_1^{(1)} + v_2^{(1)} + \ldots + v_n^{(1)} \quad (k = 0)$$

and from (3.74),

$$U_l = b_{11}^{(l)} v_1^{(1)} + b_{12}^{(l)} v_1^{(2)} + \ldots + b_{1p_1}^{(l)} v_1^{(p_1)} + b_{21}^{(l)} v_2^{(1)} + b_{22}^{(l)} v_2^{(2)} + \ldots + b_{2p_2}^{(l)} v_2^{(p_2)}$$
$$+ \ldots + b_{n1}^{(l)} v_n^{(1)} + b_{n2}^{(l)} v_n^{(2)} + \ldots + b_{np_n}^{(l)} v_n^{(p_n)}, \quad l = 1, 2, \ldots, \nu .$$

The l-th row of the matrix B has the form

$$\left(b_{11}^{(l)}, b_{12}^{(l)}, \ldots, b_{1p_1}^{(l)}, b_{21}^{(l)} b_{22}^{(l)}, \ldots, b_{2p_2}^{(l)}, \ldots, b_{n1}^{(l)}, b_{n2}^{(l)}, \ldots, b_{np_n}^{(l)}\right) .$$

Applying the operator

$$(\Delta - \mu_i)[(\Delta - \lambda_i) + (\lambda_i - \mu_i)]$$

in (3.76) to the i-th row we obtain the following recurrence formula:

$$\begin{gathered} b_{il}^{(l+1)} = (\lambda_i - \mu_l) b_{il}^{(l)}, \quad b_{ik}^{(i+1)} = (\lambda_i - \mu_l) b_{ik}^{(l)} + b_{ik-1}^{(l)}, \\ i = 1, \ldots, n; \quad k = 2, 3, \ldots, p_i. \end{gathered} \tag{3.77}$$

As follows from (3.77), the matrix B is triangular with non-zero diagonal entries $b_{ii} = 1, i = 1, 2, \ldots, p_1$,

$$b_{ii} = \prod_{j=1}^{k-1}(\lambda_k - \lambda_j)^{p_j}, \quad p_1 + p_2 + \ldots + p_{k-1} \le i \le p_1 + p_2 + \ldots + p_k.$$

To obtain the boundary values ψ_{il}, we take the trace of (3.76) on the boundary and then take the inverse of the matrix B (this procedure is not difficult because B is triangular: $\psi = B^{-1}\Psi$). The original problem is thus reduced to a boundary value problem of the type

$$\begin{aligned} &(\Delta - \lambda)^p u = 0, \quad x \in G\,, \\ &u\,|_\Gamma = \psi_1, \quad (\Delta - \lambda)u|_\Gamma = \psi_2, \ldots, (\Delta - \lambda)^{p-1}u|_\Gamma = \psi_p \end{aligned} \tag{3.78}$$

for which the mean value relation (3.67) was obtained above.

In conclusion we note that the generalization of the converse mean value relation for the problem (3.70),(3.72) is completely the same as in the case of the polyharmonic equation (keeping in mind Proposition 2.2 of Section 2.2).

Chapter 4

Triangular systems of elliptic equations

4.1. A one-component diffusion system

Let us start with the following system (which will generate nonisotropic distribution on the sphere, see Chapter 8). We seek the regular solution to the boundary value problem in a bounded domain $G \subset \mathbb{R}^3$ with a boundary Γ :

$$\begin{cases} \Delta u(x) + \gamma \frac{\partial \theta}{\partial x_1}(x) = 0 \ , \\ \Delta \theta(x) = 0 \ , \\ u|_\Gamma = \varphi_1 \ , \quad \theta|_\Gamma = \varphi_2 \ . \end{cases} \tag{4.1}$$

Let $S(x,r) \subset \bar{G}$ be an arbitrary sphere and let s be a unit vector in $\mathbb{R}^3$ with components s_i, $i = 1,2,3$. We denote by Ω the unit sphere : $\Omega = \{s \ : |s| = 1\}$.

Theorem 4.1.

For each regular solution to (4.1), *the following mean value relation holds*

$$\begin{cases} u(x) = N_r u(x) + \frac{\gamma r}{8\pi} \int\limits_{\Omega} s_1 \theta(x + rs) d\Omega(s) \ , \\ \theta(x) = \ N_r \theta(x) \ . \end{cases} \tag{4.2}$$

Proof. Since $\Delta^2 u(x) = 0$, from (3.11) (see Chapter 3) we get

$$u(x) = N_r u(x) - \frac{r^2}{6} \Delta u(x) = N_r u(x) + \frac{r^2}{6} \gamma \frac{\partial \theta}{\partial x_1}(x) \ . \tag{4.3}$$

Since $\frac{\partial \theta}{\partial x_1}(\cdot)$ is harmonic, by the volume mean value relation

$$\frac{\partial \theta}{\partial x_1}(x) = \frac{3}{4\pi r^3} \int\limits_{B(x,r)} \frac{\partial \theta}{\partial x_1}(y)dy \quad ,$$

so that, integrating by parts, one gets

$$\frac{\partial \theta}{\partial x_1}(x) = \frac{3}{4\pi r} \int\limits_{\Omega} s_1 \theta(x + rs) d\Omega(s) = \frac{3}{r} N_r(s_1 \theta) \quad . \tag{4.4}$$

From (4.3) and (4.4) we get (4.2). □

We prove now that the converse mean value relation holds.

Proposition 4.1. Integral formulation for the system.

Suppose that (4.1) *is uniquely solvable for each continuous* φ_1 *and* φ_2*; fix such functions and assume that functions* $\tilde{u}(x)$ *and* $\tilde{\theta}(x)$ *continuous in* $\bar{G}$ *with* $\tilde{u}|_\Gamma = \varphi_1$, $\tilde{\theta}|_\Gamma = \varphi_2$ *satisfy* (4.2) *at each point* $x \in G$ *for at least one sphere* $S(x, r_x) \subset G \bigcup \Gamma$*. Then the functions* $\tilde{u}$, $\tilde{\theta}$ *solve the problem* (4.1).

Proof. From Theorem 2.5 we get that $\tilde{\theta}$ is harmonic in G, thus it coincides with θ of (4.1). In view of (4.2), it holds that

$$(u - \tilde{u})(x) = N_r(u - \tilde{u})(x) + \frac{\gamma r}{8\pi} \int\limits_{\Omega} s_1(\theta - \tilde{\theta})(x + rs) d\Omega(s) \quad ,$$

which implies that $u - \tilde{u}$ is harmonic in G (again we apply Theorem 2.5). But $u|_\Gamma = \tilde{u}|_\Gamma = \varphi_1$; thus $u \equiv \tilde{u}$ in G. □

Using this Proposition we can write down a system of integral equations equivalent to the problem (4.1). Let $v = (u, \theta)^T$. Let us consider the following integral equation constructed analogous to (2.15):

$$v = \mathbf{K}_\varepsilon v + f_\varepsilon,$$

where the matrix-integral operator $\mathbf{K}_\varepsilon$ is defined by the matrix-kernel

$$k_\varepsilon(x,y) = \begin{cases} \begin{pmatrix} \delta_x(y) & \frac{\gamma r s_i}{2}\delta_x(y) \\ 0 & \delta_x(y) \end{pmatrix} & , \quad x \in G\setminus\Gamma_\varepsilon \\ 0 & , \quad x \in \Gamma_\varepsilon, \end{cases}$$

and

$$f_\varepsilon(x) = \begin{cases} \begin{pmatrix} u(x) \\ \theta(x) \end{pmatrix} & , \quad x \in G\setminus\Gamma_\varepsilon \\ 0 & , \quad x \in \Gamma_\varepsilon. \end{cases}$$

Theorem 4.2.

The spectral radius of the operator $\mathbf{K}_\varepsilon$ *is less than 1, if* $\varepsilon > 0$.

Proof. Following the proof scheme of Theorem 2.6 we find

$$||\mathbf{K}_\varepsilon^n||_{L_\infty} \le [1-\nu(\varepsilon]^{n-1}\left(1+\frac{\gamma n d^*}{2}\right)$$

which proves the theorem. □

4.2. A two-component diffusion system

For fixed integers i and j equal to 1, 2 or 3, we consider the following boundary value problem in a domain $G \subset \mathbb{R}^3$:

$$\Delta u(x) + \gamma\frac{\partial^2\theta}{\partial x_i \partial x_j}(x) = 0 \ ,$$

$$\Delta\theta(x) = 0 \ , \tag{4.5}$$

$$u|_\Gamma = \varphi_1 \ , \quad \theta|_\Gamma = \varphi_2 \ .$$

Theorem 4.3.

Each regular solution to (4.5) *satisfies the following mean value relation:*

$$u(x) = N_r u(x) + \tfrac{5}{8\pi}\gamma \int\limits_{\Omega} [s_i s_j - \tfrac{\delta_{ij}}{3}]\theta(x + rs)d\Omega(s) \quad ,$$
$$\Delta\theta(x) = N_r\theta(x) \quad , \tag{4.6}$$

where δ_{ij} is the Kronecker symbol.

Proof. From Lemma 4.1 which follows, we prove that the regular harmonic functions satisfy the relation:

$$\frac{\partial^2\theta}{\partial x_i \partial x_j}(x) = \frac{15}{4\pi r^2}\int\limits_{\Omega}[s_i s_j - \frac{\delta_{ij}}{3}]\theta(x + rs)d\Omega(s) \quad . \tag{4.7}$$

Substituting (4.7) into the mean value relation for the biharmonic function $u(x)$, one gets

$$u(x) = N_r u(x) - \frac{r^2}{6}\Delta u(x) = N_r u(x) + \frac{r^2\gamma}{6}\frac{\partial^2\theta}{\partial x_i \partial x_j}(x) \quad ,$$

and thus (4.6). □

Lemma 4.1.

Each analytical function θ defined in a bounded domain $G \subset \mathbb{R}^3$ satisfies the relations (i,j=1,2,3):

$$\frac{1}{4\pi}\int\limits_{\Omega}(a+bs_i^2)\theta(x+rs)d\Omega(s) = (a+\frac{b}{3})\theta(x)+\sum_{k=1}^{\infty}\frac{r^{2k}F_k(\theta)}{(2k+1)!(2k+3)}, \tag{4.8}$$

where

$$F_k(\theta) = a(2k+3)\Delta^k\theta(x) + b\Delta^k\theta(x) + 2kb\Delta^{k-1}\frac{\partial^2\theta}{\partial x_i^2}(x) \ ,$$

and

$$\frac{1}{4\pi}\int\limits_{\Omega} s_i s_j \theta(x + rs)d\Omega = \sum_{k=1}^{\infty}\frac{2kr^{2k}}{(2k+1)!(2k+3)}\frac{\partial^2}{\partial x_i \partial x_j}\Delta^{k-1}\theta(x) \tag{4.9}$$

for $i \neq j$.

Proof. We integrate the expansion

$$\theta(x+rs) = \theta(x) + r\sum_{k=1}^{3} \frac{\partial\theta}{\partial x_k}(x)\ s_k + \frac{r^2}{2!}\sum_{j,k=1}^{3} \frac{\partial^2\theta}{\partial x_k \partial x_j}(x)\ s_k s_j + \dots$$

over the sphere $S(x,r)$ with the weights $(a+bs_i^2)$ and $s_i s_j$.

Then the factor of the term $r^{2k}/(2k)!$ (all odd terms are, obviously, equal to zero) is

$$\sum_{A_k^\alpha} C(2\alpha_1, 2\alpha_2, 2\alpha_3) B_\alpha \frac{\partial^{2k}}{\partial x_1^{2\alpha_1} \partial x_2^{2\alpha_2} \partial x_3^{2\alpha_3}} ,$$

where

$$B_\alpha = \frac{1}{4\pi}\int_\Omega s_1^{2\alpha_1} s_2^{2\alpha_2} s_3^{2\alpha_3} d\Omega(s)$$

and the set of indices is defined as $A_k^\alpha = \{\alpha :\ \alpha_1+\alpha_2+\alpha_3 = k\}$. The structure of the coefficients $C(2\alpha_1, 2\alpha_2, 2\alpha_3)$ is clear (see Chapter 5). Using the relations

$$B_\alpha = \frac{(2\alpha_1-1)!!\,(2\alpha_2-1)!!\,(2\alpha_3-1)!!}{[2(\alpha_1+\alpha_2+\alpha_3)+1]!!}$$

and

$$\Delta^k = \sum_{A_k^\alpha} C(\alpha_1, \alpha_2, \alpha_3) \frac{\partial^{2k}}{\partial x_1^{2\alpha_1} \partial x_2^{2\alpha_2} \partial x_3^{2\alpha_3}} ,$$

where $\alpha_i = 0, 1, \dots,\ i = 1, 2, 3$, we obtain (4.8) and (4.9). □

Note that for harmonic functions the relation (4.7) follows.

Proposition 4.2. Integral formulation.

Suppose that the problem (4.5) *is uniquely solvable for each continuous* φ_1, φ_2. *Fix two such functions. Let* u *and* θ *in* $C(G\bigcup\Gamma)$ *satisfy the mean value relation* (4.6) *at each point* $x \in G$ *at least for one sphere* $S(x, r_x) \subset G$; *then they solve the problem* (4.5).

The proof repeats the considerations of Proposition 4.1.

4.3. A coupled biharmonic-harmonic equation

The results of the preceding section can be generalized for the case when the operator Δ^k acts on u. Let us treat the case $k = 2$.

We fix an integer i equal to 1, 2 or 3 and consider the following boundary value problem

$$\Delta^2 u(x) + \gamma \tfrac{\partial\theta}{\partial x_i}(x) = 0 \ , \quad x \in G \subset \mathbb{R}^3 \ ,$$

$$\Delta\theta(x) = 0 \ , \tag{4.10}$$

$$u|_\Gamma = \varphi_1 \ , \quad \Delta u|_\Gamma = \varphi_2 \ , \quad \theta|_\Gamma = \varphi_3 \ .$$

Theorem 4.4.

The regular solutions to (4.10) *satisfy the mean value relation:*

$$u(x) = N_r u(x) - \tfrac{r^2}{6} N_r \Delta u(x) - \tfrac{7r^3}{120}\gamma N_r(s_i\theta)(x) \ ,$$

$$\Delta u(x) = N_r \Delta u(x) + \tfrac{\gamma r}{2} N_r(s_i\theta)(x) \ ,$$

$$\theta(x) = N_r\theta(x) \ .$$

Proof. Since $\Delta^3 u(x) = 0$, from (3.10) we find that

$$u(x) = N_r u(x) - \tfrac{r^2}{6}\Delta u(x) - \tfrac{r^4}{120}\Delta^2 u(x) \ ,$$

$$\Delta u(x) = N_r(\Delta u)(x) - \tfrac{r^2}{6}\Delta^2 u(x) \ ,$$

$$\theta(x) = N_r\theta(x) \ .$$

Thus,

$$u(x) = N_r u(x) - \tfrac{r^2}{6} N_r(\Delta u)(x) - \tfrac{7r^4}{360}\gamma\tfrac{\partial\theta}{\partial x_i}(x) \ ,$$

$$\Delta u(x) = N_r(\Delta u)(x) + \tfrac{\gamma r^2}{6}\tfrac{\partial\theta}{\partial x_i}(x) \ ,$$

$$\theta(x) = N_r\theta(x) \ ,$$

which, in view of (4.4), gives the desired result. □

Proposition 4.3. Integral Formulation.

Suppose that the problem (4.10) *is uniquely solvable for all continuous functions* $\varphi_1, \varphi_2, \varphi_3$.

Let $\varphi_1, \varphi_2, \varphi_3$ *be fixed. Let* $v := (v_1, v_2, v_3)^T$, $v_i \in C(G \bigcup \Gamma)$ *satisfy the mean value relation*

$$v(x) = N_r(Av) \quad ,$$

where

$$A = \begin{pmatrix} 1 & -\frac{r^2}{6} & -\frac{7\gamma}{120} r^3 s_i \\ 0 & 1 & \frac{\gamma r}{2} \\ 0 & 0 & 1 \end{pmatrix}$$

for each $x \in G$, *at least for one sphere* $S(x, r_x) \subset \bar{G}$. *Suppose also that*

$$v_1|_\Gamma = \varphi_1 \ , \quad v_2|_\Gamma = \varphi_2 \ , \quad v_3|_\Gamma = \varphi_3 \quad .$$

Then v *solves the problem* (4.10).

The proof repeats the arguments given in Proposition 4.1.

In conclusion we give the following extension.

Let us consider the problem:

$$\begin{aligned} &\Delta^2 u(x) + \gamma \frac{\partial^2 \theta}{\partial x_i \partial x_j}(x) = 0 \quad , \quad x \in G \subset \mathbb{R}^3 \quad , \\ &\Delta \theta(x) = 0 \quad , \\ &u|_\Gamma = \varphi_1 \ , \quad \Delta u|_\Gamma = \varphi_2 \ , \quad \theta|_\Gamma = \varphi_3 \quad . \end{aligned} \tag{4.11}$$

The above integral formulation is true here with the matrix

$$A = \begin{pmatrix} 1 & -\frac{r^2}{6} & -\frac{7\gamma r^2}{24}(s_i s_j - \frac{\delta_{ij}}{3}) \\ 0 & 1 & \frac{5\gamma}{2}(s_i s_j - \frac{\delta_{ij}}{3}) \\ 0 & 0 & 1 \end{pmatrix} \quad .$$

Chapter 5

Systems of elasticity theory

5.1. The Lamé equation

5.1.1. Direct spherical mean value theorem

Suppose a homogeneous isotropic medium $G \subset \mathbb{R}^n$ with a boundary Γ is given, whose state in the absence of body forces is governed by the classical static equation, the Lamé equation:

$$\Delta \mathbf{u}(x) + \alpha \operatorname{grad} \operatorname{div} \mathbf{u}(x) = 0, \quad x \in G, \tag{5.1}$$

where $\mathbf{u}(x) = (u_1(x_1, \ldots, x_n), \ldots, u_n(x_1, \ldots, x_n))$ is a vector of displacements, whose components are real-valued regular functions. The elastic constant α

$$\alpha = \frac{\lambda + \mu}{\mu} = \frac{1}{1 - 2\sigma}$$

is expressed through the Lamé constants of elasticity λ and μ, and σ, the Poisson ratio (for $-1 < \sigma < 1/2$ we have $1/3 < \alpha$). Here we will employ the summation convention, for example,

$$u_{i,jj} = \sum_{j=1}^{n} \frac{\partial^2 u_i}{\partial x_j^2} \; .$$

Hence, the equation (5.1) can be written in the form

$$u_{i,jj} + \alpha\, u_{j,ji} = 0, \quad i, j = 1, \ldots, n \quad . \tag{5.2}$$

The first boundary value problem for the Lamé equation consists in finding a vector function $\mathbf{u} \in C^2(G) \cap C(\bar{G})$ satisfying the boundary condition

$$\mathbf{u}(y) = \mathbf{g}(y), \quad y \in \Gamma \ , \tag{5.3}$$

where $\mathbf{g} \in C(\Gamma)$ is a given vector-function.

We first give the direct spherical mean value relation carried out in [6].

In addition to the mean value theorems for the biharmonic functions proved above, the following spherical mean value relation is deduced.

Lemma 5.1.

For any function $v(x_1, \dots, x_n)$ biharmonic in $G \subset \mathbb{R}^n$ the following relation is valid for arbitrary ball $B(x, r) \subset G$

$$v(0) = \frac{n}{2\omega_n}\Big[\frac{n+2}{R^n} \int\limits_{B(0,R)} v \, dV - \frac{1}{R^{n-1}} \int\limits_{S(0,R)} v \, dS\Big], \tag{5.4}$$

where $\omega_n = 2\pi^{n/2}/\Gamma(n/2)$ is the area of the surface of the unit sphere in $\mathbb{R}^n$.

Proof. The function Δv is harmonic; therefore

$$\Delta v(0) = \frac{n}{\omega_n r^n} \int\limits_{B(0,r)} \Delta v \, dV = \frac{n}{\omega_n r^n} \int\limits_{S(0,r)} \frac{\partial v}{\partial r} \, dS$$
$$= \frac{n}{\omega_n r} \int\limits_{\Omega} \frac{\partial v}{\partial r} \, d\Omega.$$

Thus

$$r \Delta v(0) = \frac{n}{\omega_n} \int\limits_{\Omega} \frac{\partial v}{\partial r} \, d\Omega, \tag{5.5}$$

and the integration from $r = 0$ to $r = \rho > 0$ (interchanging the order of integration on the right-hand side) yields

$$\frac{\rho^2}{2} \Delta v(0) = \frac{n}{\omega_n}\Big[\int\limits_{\Omega} v \, d\Omega - \omega_n v(0)\Big] \ .$$

Replacing $\Delta v(0)$ by its value from (5.5) we get

$$\rho^n \int_\Omega \frac{\partial v}{\partial \rho}\, d\Omega = 2\rho^{n-1}\Big[\int_\Omega v\, d\Omega - \omega_n v(0)\Big].$$

By adding $n\rho^{n-1}\int_\Omega v\, d\Omega$ to both sides we come to

$$\int_\Omega \frac{\partial[\rho^n v]}{\partial \rho}\, d\Omega = \rho^{n-1}\Big[(n+2)\int_\Omega v\, d\Omega - 2\omega_n v(0)\Big]. \tag{5.6}$$

Integrating (5.6) with respect to ρ from $\rho = 0$ to $\rho = R$ yields

$$R^n \int_\Omega v\, d\Omega = (n+2)\int_{B(0,R)} v\, dV - 2\frac{\omega_n R^n}{n} v(0),$$

from which the desired result follows. □

Theorem 5.1.

The regular solutions to the system (5.2) *satisfy the following spherical mean value relation*

$$u_i(0) = \frac{n}{2(n+\alpha)\omega_n}\int_\Omega \Big((2-\alpha)\delta_{ij} + (n+2)\alpha\frac{x_i x_j}{R^2}\Big) u_j\, d\Omega, \tag{5.7}$$

$i = 1, \ldots, n,$ δ_{ij} *is the Kronecker symbol.*

Proof. It is easy to show that the displacement components u_i satisfying (5.2) are biharmonic functions and the divergence of $\mathbf{u}$ is harmonic. Indeed, (5.2) yields

$$u_{i,jji} + \alpha\, u_{j,jii} = 0,$$

or

$$(1+\alpha)\Delta u_{j,j} = 0.$$

Next, applying the operator Δ to (5.2) we get

$$0 = \Delta[u_{i,jj} + \alpha u_{j,ji}] = \Delta^2 u_i + \alpha(\Delta u_{j,j})_{,i}\ ,$$

or

$$\Delta^2 u_i = 0. \tag{5.8}$$

We use now the fact that u_i are biharmonic and that the mean value theorem of Lemma 5.1 is true to derive the spherical mean value relation for the displacements u_i.

We integrate by parts the Lamé equation

$$\begin{aligned} 0 &= \int\limits_{B(0,R)} r^2[u_{i,jj} + \alpha u_{j,ji}]\, dV \\ &= \int\limits_{S(0,R)} R^2\left[u_{i,j}\frac{x_j}{R} + \alpha u_{j,j}\frac{x_i}{R}\right] dS - 2\int\limits_{B(0,R)} [x_j u_{i,j} + \alpha x_i u_{j,j}]\, dV, \end{aligned} \tag{5.9}$$

where $x_i/R = s_i$ is the i–th component of the direction cosines of the vector $\mathbf{s}$. The surface integral in (5.9) is zero since

$$\begin{aligned} &\int\limits_{S(0,R)} R^2\left[u_{i,j}\frac{x_j}{R} + \alpha u_{j,j}\frac{x_i}{R}\right] dS \\ &\qquad = R^2 \int\limits_{B(0,R)} [u_{i,jj} + \alpha u_{j,ji}]\, dV = 0. \end{aligned}$$

Integrating the remaining integral in (5.9) by parts yields

$$\begin{aligned} 0 = &\int\limits_{S(0,R)} \left[R u_i + \alpha\frac{x_i x_j}{R} u_j\right] dS \\ &- \int\limits_{B(0,R)} (n+\alpha) u_i\, dV. \end{aligned}$$

Therefore, (since $dS = R^{n-1} d\Omega$)

$$\int\limits_{B(0,R)} u_i\, dV = \frac{R^n}{n+\alpha}\int\limits_{\Omega} \left[\left(\delta_{ij} + \alpha\frac{x_i x_j}{R^2}\right) u_j\right] d\Omega. \tag{5.10}$$

Applying the mean value theorem for biharmonic functions (5.4) to the displacement u_i, and then omitting the integral over $B(0,R)$ we obtain the desired result. □

Note that the direct spherical mean value relation (5.7) can be derived from the Betty formula (e.g. see [25]). This formula in $\mathbb{R}^3$ has the form

$$\int_G (\mathbf{v} \cdot A\mathbf{u} - \mathbf{u} \cdot A\mathbf{v})\, dV = \int_{\partial G} (\mathbf{v} \cdot t(\mathbf{u}) - \mathbf{u} \cdot t(\mathbf{v}))\, dS \ . \tag{5.11}$$

Here $A = \Delta^*$, and $t(\mathbf{v})$ is the stress vector:

$$t(\mathbf{v}) = \sum_{i,k=1}^{3} \tau_{ik} \cos(n, x_i)\mathbf{e}_k,$$

where τ_{ik} are the components of the stress tensor:

$$\tau_{ik} = \mu \left(\frac{\partial u_i}{\partial x_k} + \frac{\partial u_k}{\partial x_i} \right) + \lambda \delta_{ik} \operatorname{div} \mathbf{u} \ ,$$

and $\mathbf{e}_k$ is the ort of the axis x_k.

Assume that the function $\mathbf{u}$ is twice continuously differentiable in G and belongs to $C(\bar{G})$. Let us cut off a sphere $S(x, \varepsilon) \subset G$, ε being small enough. We apply the Betty formula (5.11) to the domain $G \setminus S(x, \varepsilon)$ and take the limit as $\varepsilon \to 0$. Taking into account the relations

$$\lim_{\varepsilon \to 0} \int_{G \setminus S(x,\varepsilon)} \mathbf{v}_i \cdot A\mathbf{u}\, dV = \int_G \mathbf{v}_i \cdot A\mathbf{u}\, dV \ ,$$

$$\lim_{\varepsilon \to 0} \int_{S(x,\varepsilon)} \mathbf{v}_i \cdot t(\mathbf{u})\, dS \ = 0 \ ,$$

$$\lim_{\varepsilon \to 0} \int_{S(x,\varepsilon)} \mathbf{u} \cdot t(\mathbf{v}_i)\, dS \ = -u_i(x)$$

we obtain

$$u_i(x) = \int_G \mathbf{v}_i \cdot A\mathbf{u}\, dV - \int_{\partial G} \{\mathbf{u} \cdot t(\mathbf{v}_i) - \mathbf{v}_i \cdot t(\mathbf{u})\}\, dS \ . \tag{5.12}$$

Here $\mathbf{v}_i = (v_{i1}, v_{i2}, v_{i3})^T$, v_{ik} being the components of the tensor of fundamental solutions $\mathcal{V}$:

$$v_{ik} = \frac{1}{16\pi\mu(1-\sigma)}\left\{(3-4\sigma)\frac{\delta_{ik}}{r} + \frac{(x_i-y_i)(x_k-y_k)}{r^3}\right\} .$$

Multiplying (5.12) by $\mathbf{e}_i$ and summing over i we find

$$\mathbf{u}(x) = \int_G \mathcal{V}\cdot A\mathbf{u}\,dV - \int_{\partial G}\{T(\mathcal{V})\cdot\mathbf{u} - \mathcal{V}\cdot t(\mathbf{u})\}\,dS\ , \tag{5.13}$$

where $T(\mathcal{V})$ is a tensor whose ith row coincides with $t(\mathbf{v}_i)$.

In our case, G is the ball $B(x,R)$; therefore

$$T_{ik}(\mathcal{V}) = \frac{1-2\sigma}{8\pi(\sigma-1)R^2}\left[\delta_{ik} + \frac{3}{1-2\sigma}\frac{(x_i-y_i)(x_k-y_k)}{R^2}\right] .$$

Let us define a tensor $\hat{\mathcal{V}}$ by

$$\hat{v}_{ik} = \frac{1}{16\pi\mu(1-\sigma)}\left\{(3-4\sigma)\frac{\delta_{ik}}{R} + \frac{(x_i-y_i)(x_k-y_k)}{R^3}\right\} .$$

Note that $\mathcal{V} = \hat{\mathcal{V}}$ for $r = R$. Applying the Betty formula for $\hat{\mathcal{V}}$ we get

$$\int_{S(x,R)} \mathcal{V}\cdot t(\mathbf{u})\,dS = \int_{S(x,R)} \hat{\mathcal{V}}\cdot t(\mathbf{u})\,dS = \int_{S(x,R)} T(\hat{\mathcal{V}})\cdot\mathbf{u}\,dS$$
$$+ \int_{B(x,R)} \{\hat{\mathcal{V}}\cdot A\mathbf{u} - \mathbf{u}\cdot A\hat{\mathcal{V}}\}\,dV .$$

Substituting this representation into (5.13) we come to the following mean value relation

$$\begin{aligned}\mathbf{u}(x) = \int_{B(x,R)} (\mathcal{V}+\hat{\mathcal{V}})\cdot A\mathbf{u}\,dV - \int_{B(x,R)} A\hat{\mathcal{V}}\cdot\mathbf{u}\,dV \\ - \int_{S(x,R)} \{T(\mathcal{V}) - T(\hat{\mathcal{V}})\}\cdot\mathbf{u}\,dS\ .\end{aligned} \tag{5.14}$$

Here

$$(A\hat{\mathcal{V}})_{ik} = \frac{2\lambda+3\mu}{8\pi\mu(1-\sigma)R^3}\delta_{ik}\ ,$$

and

$$(T(\hat{\mathcal{V}}))_{ik} = \frac{1}{16\pi\mu(1-\sigma)R^4}\{\mu R^2\delta_{ik} + (4\lambda+3\mu)(x_i-y_i)(x_k-y_k)\} .$$

In the case when $\mathbf{u}$ is the solution to the Lamé equation we have $A\mathbf{u} = 0$; hence

$$\mathbf{u}(x) = -\int\limits_{B(x,R)} A\hat{\mathcal{V}}\cdot\mathbf{u}\,dV - \int\limits_{S(x,R)} \{T(\mathcal{V}) - T(\hat{\mathcal{V}})\}\cdot\mathbf{u}\,dS\ , \tag{5.15}$$

or, more explicitly,

$$u_i(x) = \frac{2\lambda+3\mu}{8\pi R^3\mu(\sigma-1)}\int\limits_{B(x,R)} u_i\,dV$$
$$-\frac{1}{16\pi R^2(\sigma-1)}\int\limits_{S(x,R)}\left\{(3-4\sigma)\delta_{ik} + \left(9+4\frac{\lambda}{\mu}\right)\frac{(x_i-y_i)(x_k-y_k)}{R^2}\right\}u_k dS.$$

Note that if we change the volume integral in (5.15) with the representation (5.10) we obtain the spherical mean value relation (5.7).

5.1.2. Converse spherical mean value theorem

As in the case of harmonic functions, it is not difficult to show that the weak converse spherical mean value relation holds also for the regular solutions to the Lamé equation (e.g. see [5]). It can be derived, for instance, from the series expansion of the spherical mean with respect to a vector measure (see the expansion (5.21)). However the strong converse spherical mean value relation can be proved by using the approach [5] and the converse mean value relation of Theorem 2.5 (see [41]).

Theorem 5.2. Integral Formulation.

Suppose that the boundary Γ *of* $G \subset \mathbb{R}^n$ *is regular so that the Dirichlet problem* (5.1),(5.3) *is uniquely solvable for arbitrary continuous boundary functions* $g_1, \ldots, g_n$, *and the solutions have third continuous derivatives in* $G \cup \Gamma$. *Assume that a vector function* $\mathbf{u} \in C^3(G\cup\Gamma)$, $(\mathbf{u}|_\Gamma = \mathbf{g})$ *satisfies the mean value relation* (5.7) *for each* $x \in G$ *at least for one sphere* $S(x, r_x) \subset \bar{G}$. *Then* $\mathbf{u}(\cdot)$ *solves the Dirichlet problem* (5.1),(5.3).

Proof. Take an arbitrary sphere $S(y,r) \subset \bar{G}$, and set for an arbitrary point $x \in B(y,r)$

$$\begin{aligned}\Theta_i(x) := u_i(x) &+ \frac{2-\alpha}{2(n+\alpha)}(x_j - y_j)(u_{i,j}(x) - u_{j,i}(x)) \\ &+ \frac{(n+1)\alpha+2}{2(n+\alpha)}(x_i - y_i)u_{j,j}(x) \quad .\end{aligned} \tag{5.16}$$

Note that the Lamé equation can be rewritten in the form: $\Delta\Theta(x) = 0$. Indeed, we first differentiate the Lamé equation. Using the properties

$$\Delta u_{j,j} = 0, \quad \Delta(u_{i,j} - u_{j,i}) = 0$$

we find by differentiating the Lamé equation that

$$u_{i,jj} + \frac{2-\alpha}{n+\alpha}(u_{i,jj} - u_{j,ij}) + \frac{(n+1)\alpha+2}{n+\alpha}u_{j,ji} = 0 \ .$$

This yields

$$\frac{n+2}{n+\alpha}\Delta u_i + \frac{\alpha(n+2)}{n+\alpha}\frac{\partial}{\partial x_i}\operatorname{div}\mathbf{u} = 0$$

which coincides with the Lamé equation, since $n + \alpha > 0$. Obviously, these transformations can be converted, i.e., starting from the equation $\Theta(x) = 0$ we come back along the same transformations to the Lamé equation. Thus, the Lamé equation is equivalent to the equation $\Theta(x) = 0$.

Now, we start from the spherical mean value relation given in the theorem formulation and integrate by parts the right-hand side of the relation (5.7) on the sphere $S(y,r)$. This leads, as we will prove below, to

$$\begin{aligned}u_i(y) = \frac{1}{\omega_n r^n}\int_{B(y,r)} \Big\{ &n\, u_i(x) + a(x_j - y_j)(u_{i,j}(x) - u_{j,i}(x)) \\ &+ (a+b)(x_i - y_i)u_{j,j}(x)\Big\} dx \ ,\end{aligned} \tag{5.17}$$

where

$$a = \frac{(2-\alpha)n}{2(n+\alpha)}, \quad b = \frac{n(n+2)\alpha}{2(n+\alpha)} \ .$$

This relation implies by Theorem 2.5 that the functions Θ_i in (5.16) are harmonic in G. Thus we get $\Delta\Theta(y) = 0$ for all $y \in G$ which implies that at y, one has that

$$\Delta u_i(y) + \alpha \frac{\partial}{\partial x_i} \operatorname{div} \mathbf{u}(y) = 0 \ , \quad i = 1, \ldots, n.$$

Since y was chosen arbitrarily in G, this completes the proof. It remains only to transform the right-hand side of the relation (5.7) to the right-hand side of (5.17). However we carry out the transformations backwards, i.e., we deduce the right-hand side of (5.7) from the right-hand side of (5.17). The transformations can then be easily traced back.

Thus we integrate by parts the right-hand side of (5.17) using the formula

$$\int\limits_{B(y,r)} g(x) \frac{\partial f(x)}{\partial x_i} dx = \int\limits_{S(y,r)} f(z) g(z) n_i \, dS_z - \int\limits_{B(y,r)} f(x) \frac{\partial g(x)}{\partial x_i} dx \ .$$

This leads to the following representations:

$$a \int\limits_{B(y,r)} (x_j - y_j) u_{i,j}(x) dx = ar \int\limits_{S(y,r)} u_i(x) \, dS_x - an \int\limits_{B(y,r)} u_i(x) \, dx \ ,$$

$$a \int\limits_{B(y,r)} (x_j - y_j) u_{j,i}(x) dx = a \int\limits_{S(y,r)} \frac{(x_j - y_j)(x_i - y_i)}{r} u_j(x) dS_x - a \int\limits_{B(y,r)} u_i(x) \, dx \ ,$$

and

$$(a+b) \int\limits_{B(y,r)} (x_i - y_i) \operatorname{div} \mathbf{u}(x) dx = (a+b) \int\limits_{S(y,r)} \frac{(x_i - y_i)(x_j - y_j)}{r} u_j(x) \, dS_x - (a+b) \int\limits_{B(y,r)} u_i(x) \, dx \ .$$

We now evaluate the right-hand side of (5.17) using these representations. The volume integrals disappear, since $an + b = n$. Thus we finally get

$$u_i(y) = \frac{1}{\omega_n r^{n-1}} \left\{ a \int\limits_{S(y,r)} u_i dS_x + b \int\limits_{S(y,r)} \frac{(x_i - y_i)(x_j - y_j)}{r^2} u_j \, dS_x \right\} \quad (5.18)$$

which is the desired spherical mean value relation coinciding with (5.7). As mentioned above, these transformations can be traced backwards to deduce (5.17) from (5.7), which proves the theorem. □

5.2. Pseudo–vibration elastic equation

Let us define the operator Δ^* by

$$\Delta^* = \mu\Delta + (\lambda + \mu)\,\text{grad div} \ .$$

Consider the pseudo-vibration elastic equation:

$$\Delta^*\mathbf{u}(x) - \nu\mathbf{u}(x) = 0, \quad x \in G \subset \mathbb{R}^n \ , \tag{5.19}$$

where $\mathbf{u} = (u_1(x), \ldots, u_n(x))$ is a real-valued regular vector function; ν is a complex parameter.

Now we obtain the mean value relation for this equation. In accordance with the scheme described in Section 2.1.1, we introduce an averaging measure over a sphere $S(x,r)$:

$$d\eta_1(s) = \left\{\frac{1}{\omega_n r^{n-1}}\left[a\delta_{ij} + bs_is_j\right]\, dS(s)\right\}_{i,j=1}^{n},$$

where a and b are some constants, s_i is the i-th direction cosine of the vector $s \in \Omega(x,1)$, δ_{ij} is the Kronecker symbol, ω_n is the surface area of $\Omega(x,1)$.

The corresponding operator averaging the vector function $\mathbf{u}$ over the sphere $S(x,r)$ with respect to the measure $d\eta_1$ will be denoted by $N^1\mathbf{u}(x)$:

$$N^1\mathbf{u}(x) = \int\limits_{S(x,r)} \mathbf{u}(x + rs)d\eta_1(s), \tag{5.20}$$

where the integral of the vector $\mathbf{u}(x + rs)$ is understood to be a vector whose components are integrals of the corresponding elements. The integrand in (5.20) is regarded as a product of the matrix $d\eta_1(s)$ by the vector $\mathbf{u}(x + rs)$.

We now present some auxiliary results which will be used in obtaining the main result, namely, Theorem 5.7 on the mean value relation for (5.19) (see [44]).

Theorem 5.3.

For a real-valued vector function $\mathbf{u} = (u_1(x), \dots, u_n(x))^T$, $x \in G \subset \mathbb{R}^n$, *analytic in* G, *the following expansion*

$$N^1 \mathbf{u}\,(x) = \left(a + \frac{b}{n}\right)\mathbf{u}(x) + \sum_{k=1}^{\infty} \frac{r^{2k}(n-2)!!}{(2k)!!(n+2k)!!} \times \left\{(n+2k)a\Delta^k \mathbf{u}(x) + b\Delta^k \mathbf{u}(x) + 2kb\Delta^{k-1}\operatorname{grad}\operatorname{div}\,\mathbf{u}(x)\right\} \tag{5.21}$$

holds if the series on the right-hand side converges.

Proof. For simplicity, consider the case $n = 3$. The proof for arbitrary dimension is analogous. A scalar case was discussed briefly in Section 4.2, Lemma 4.1. Consider the first relation of the system (5.21). By definition of N^1 (see (5.20)) we have:

$$\left(N^1\mathbf{u}\right)_1(x) = \frac{1}{4\pi r^2}\int\limits_{S(x,r)} (a + bs_1^2)u_1(x+rs)\,dS + \frac{b}{4\pi r^2}\int\limits_{S(x,r)} s_1 s_2 u_2(x+rs)\,dS + \frac{b}{4\pi r^2}\int\limits_{S(x,r)} s_1 s_3 u_3(x+rs)\,dS. \tag{5.22}$$

Substituting the Taylor series expansion of the analytic function u_1

$$u_1(x+rs) = u_1(x) + r\sum_{k=1}^{3} \frac{\partial u_1(x)}{\partial x_k} s_k + \frac{r^2}{2!}\sum_{j,k=1}^{3} \frac{\partial^2 u_1(x)}{\partial x_k \partial x_j} s_k s_j + \dots.$$

into the first term of (5.22), then integrating this series, term by term, we find

$$\frac{1}{4\pi r^2}\int\limits_{S(x,r)} (a + bs_1^2)u_1(x+rs)\,dS = \sum_{k=0}^{\infty} \frac{r^{2k}}{(2k)!} \sum_{A_\alpha^k} C(2\alpha_1, 2\alpha_2, 2\alpha_3) \frac{\partial^{2k} u_1(x)}{\partial x_1^{2\alpha_1}\partial x_2^{2\alpha_2}\partial x_3^{2\alpha_3}} \times \frac{1}{4\pi r^2}\int\limits_{S(x,r)} (a + bs_1^2) s_1^{2\alpha_1} s_2^{2\alpha_2} s_3^{2\alpha_3}\,dS, \tag{5.23}$$

where the set of indices is defined as $A_\alpha^k = \{(\alpha_1, \alpha_2, \alpha_3) : \alpha_1+\alpha_2+\alpha_3 = k\}$, and

$$C(l,n,m) = \frac{(l+n+m)!}{l!\,n!\,m!}.$$

Taking into account the relations

$$\int\limits_{\Omega(x,1)} s_1^{2\alpha_1} s_2^{2\alpha_2} s_3^{2\alpha_3}\, d\Omega = 4\pi \frac{(2\alpha_1-1)!!(2\alpha_2-1)!!(2\alpha_3-1)!!}{((2\alpha_1+2\alpha_2+2\alpha_3)+1)!!},$$

and

$$\Delta^k = \sum_{A_\alpha^k} C(\alpha_1,\alpha_2,\alpha_3) \frac{\partial^{2k}}{\partial x_1^{2\alpha_1} \partial x_2^{2\alpha_2} \partial x_3^{2\alpha_3}},$$

we obtain

$$\begin{aligned} \frac{1}{4\pi} \int\limits_{S(x,r)} (a+bs_1^2)u_1(x+rs)\, dS &= \Big(a+\frac{b}{3}\Big)u_1(x) \\ &+ \sum_{k=1}^{\infty} \frac{r^{2k}}{(2k+1)!(2k+3)} \Big\{a(2k+3)\Delta^k u_1(x) + bBu_1(x)\Big\}, \end{aligned} \tag{5.24}$$

where

$$Bu_1(x) = \sum_{A_\alpha^k} C(2\alpha_1, 2\alpha_2, 2\alpha_3)(2\alpha_1+1) \frac{\partial^{2k} u_1(x)}{\partial x_1^{2\alpha_1} \partial x_2^{2\alpha_2} \partial x_3^{2\alpha_3}}.$$

We transform the operator B. To this end, we introduce an auxiliary operator F dependent on a parameter γ

$$F_\gamma u_1(x) = \sum_{A_\alpha^k} C(2\alpha_1, 2\alpha_2, 2\alpha_3)\gamma^{(2\alpha_1+1)} \frac{\partial^{2k} u_1(x)}{\partial x_1^{2\alpha_1} \partial x_2^{2\alpha_2} \partial x_3^{2\alpha_3}},$$

such that

$$B = F'_\gamma\Big|_{\gamma=1}.$$

But

$$F_\gamma u_1(x) = \gamma \Big\{ \gamma^2 \frac{\partial^2}{\partial x_1^2} + \frac{\partial^2}{\partial x_2^2} + \frac{\partial^2}{\partial x_3^2} \Big\}^k u_1(x);$$

therefore, to get the desired expression of B, it is sufficient to differentiate the last expression with respect to γ at the point $\gamma = 1$:

$$Bu_1(x) = \Delta^k u_1(x) + 2k\Delta^{k-1}\frac{\partial^2 u_1(x)}{\partial x_1^2}.$$

Similar expansions for the second and third integral terms in (5.22) also can be derived:

$$\frac{b}{4\pi r^2}\int\limits_{S(x,r)} s_1 s_2 u_2(x+rs)\, dS = b\sum_{k=1}^{\infty} M_k(r) \times\left\{2k\frac{\partial^2}{\partial x_1 \partial x_2}\Delta^{k-1}u_2(x)\right\},$$

$$\frac{b}{4\pi r^2}\int\limits_{S(x,r)} s_1 s_3 u_3(x+rs)\, dS = b\sum_{k=1}^{\infty} M_k(r) \times\left\{2k\frac{\partial^2}{\partial x_1 \partial x_3}\Delta^{k-1}u_3(x)\right\},$$

where $M_k(r) = r^{2k}/[(2k+1)!(2k+3)]$. Then the expression (5.22) transforms to

$$\left(N^1\mathbf{u}\right)_1(x) = \sum_{k=0}^{\infty}\frac{r^{2k}}{(2k+1)!(2k+3)} \times\left\{a(2k+3)\Delta^k u_1(x) + b\Delta^k u_1(x) + 2kb\Delta^{k-1}\frac{\partial}{\partial x_1}\mathrm{div}\mathbf{u}\right\}. \tag{5.25}$$

Repeating the calculations (5.22)-(5.25) for the second and third components of the averaging operator $N^1\mathbf{u}$, we obtain the desired result (5.21) for the case $\mathbf{u} \in \mathbb{R}^3$

$$N^1\mathbf{u}(x) = \sum_{k=1}^{\infty}\frac{r^{2k}}{(2k+1)!(2k+3)}\{a(2k+3)\Delta^k\mathbf{u}(x) + b\Delta^k\mathbf{u}(x) + 2kb\Delta^{k-1}\mathrm{grad}\,\mathrm{div}\mathbf{u}(x)\} + \left(a+\frac{b}{3}\right)\mathbf{u}(x).$$

□

Corollary 5.1.

For the solution to equation (5.19) *the following expansion holds:*

$$N^1\mathbf{u}(x) - \mathbf{u}(x) = \sum_{k=1}^{\infty} \frac{r^{2k}(n-2)!!}{(2k)!!(n+2k)!!} \times \left\{ \frac{kn(n+2)\nu}{\lambda + \mu(n+1)} \Delta^{k-1}\mathbf{u}(x) - n(k-1)\Delta^k \mathbf{u}(x) \right\}. \tag{5.26}$$

Proof. Setting $\beta = (n+2)(\lambda+\mu)/2[\lambda+\mu(n+1)]$, $a = 1-\beta$, $b = n\beta$ in (5.21), for the solution to the equation (5.19) we obtain (5.26). □

Corollary 5.2.

The mean value relation for the Lamé equation

$$\Delta^* \mathbf{u}(x) = 0$$

follows immediately from (5.26)

$$N^1\mathbf{u}(x) = \mathbf{u}(x).$$

We need certain properties of the solutions to equation (5.19). Let us formulate the following theorem (see [21], III.2.4).

Theorem 5.4.

The general solution to (5.19) *can be represented in the form* $\mathbf{u}(x) = \mathbf{u}^p(x) + \mathbf{u}^s(x)$, *where* $\mathbf{u}^p$ *and* $\mathbf{u}^s$ *are regular vector functions determined by the equations*

$$\begin{aligned} (\Delta - k_p^2)\mathbf{u}^p &= 0, \qquad \operatorname{rot} \mathbf{u}^p = 0, \\ (\Delta - k_s^2)\mathbf{u}^s &= 0, \qquad \operatorname{div} \mathbf{u}^s = 0, \end{aligned} \tag{5.27}$$

where $k_p = \sqrt{\nu/(\lambda+2\mu)}$, $k_s = \sqrt{\nu/\mu}$, *and*

$$\mathbf{u}^p = \frac{(\Delta - k_s^2)\mathbf{u}}{k_p^2 - k_s^2}, \quad \mathbf{u}^s = \frac{(\Delta - k_p^2)\mathbf{u}}{k_s^2 - k_p^2}, \quad u \in C^\infty(G). \tag{5.28}$$

Using this decomposition of $\mathbf{u}(x)$ into the potential $(\mathbf{u}^p(x))$ and solenoidal $(\mathbf{u}^s(x))$ components, we now obtain the following properties.

Lemma 5.2.

The solution to (5.19) *satisfies the relations*

$$\Delta^n \mathbf{u}(x) = k_s^{2n}\mathbf{u}^s(x) + k_p^{2n}\mathbf{u}^p(x), \tag{5.29}$$

$$\Delta^n \operatorname{div} \mathbf{u}(x) = k_p^{2n} \operatorname{div} \mathbf{u}(x). \tag{5.30}$$

Proof. The repeated operator Δ^* is represented as

$$\nu^n \mathbf{u} = \mu^n \Delta^n \mathbf{u} + \Big(\frac{\nu}{\lambda+2\mu}\Big)^{n-1} \frac{\nu\mathbf{u} - \mu\Delta\mathbf{u}}{\lambda+\mu}\{(\lambda+2\mu)^n - \mu^n\}.$$

Hence,

$$\Delta^n \mathbf{u} = \Big(\frac{\nu}{\mu}\Big)^{n-1} \frac{\lambda+2\mu}{\lambda+\mu}\Big[\Delta\mathbf{u} - \frac{\nu\mathbf{u}}{\lambda+2\mu}\Big]$$
$$- \Big(\frac{\nu}{\lambda+2\mu}\Big)^{n-1} \frac{\mu}{\lambda+\mu}\Big[\Delta\mathbf{u} - \frac{\nu\mathbf{u}}{\mu}\Big].$$

Recalling that

$$(k_s^2 - k_p^2)^{-1} = \frac{\mu(\lambda+2\mu)}{\nu(\lambda+\mu)}$$

we obtain (see Theorem 5.4)

$$\Delta^n \mathbf{u} = k_s^{2n} \frac{(\Delta - k_p^2)\mathbf{u}}{k_s^2 - k_p^2} + k_p^{2n} \frac{(\Delta - k_s^2)\mathbf{u}}{k_p^2 - k_s^2},$$

thus, (5.29) is proved in view of Theorem 5.4. The property (5.30) is directly implied by the last formula (it is sufficient to apply the operator div to both sides of (5.29) and make use of (5.27)). □

Theorem 5.5.

For any analytical functions $\mathbf{u}^p$, $\mathbf{u}^s$ *(case* $\mathbb{R}^3$*) satisfying the equations* (5.27), (5.28) *the following relation holds:*

$$N_r \mathbf{u} = \frac{\sinh(rk_p)}{rk_p}\mathbf{u}^p(x) + \frac{\sinh(rk_s)}{rk_s}\mathbf{u}^s(x). \tag{5.31}$$

Proof. In Section 2.1.1 we used the expansion of the averaging operator $N_r v$ corresponding to the uniform measure $d\Omega$. Let us complete this definition for the case of vector function $\mathbf{v}$:

$$(N_r\mathbf{v})_i = N_r(v_i) = \sum_{n=0}^{\infty} \frac{r^{2n}}{(2n+1)!}\Delta^n v_i(x), \quad i = 1, 2, 3.$$

Then, by definition and on the basis of (5.29) we have

$$N_r\mathbf{u}(x) = \sum_{n=0}^{\infty} \frac{r^{2n}}{(2n+1)!}\{k_s^{2n}\mathbf{u}^s(x) + k_p^{2n}\mathbf{u}^p(x)\}$$

$$= \frac{\sinh(rk_s)}{rk_s}\mathbf{u}^s(x) + \frac{\sinh(rk_p)}{rk_p}\mathbf{u}^p(x).$$

□

Theorem 5.6.

For the solution to the equation (5.19) *(case* $\mathbb{R}^3$*) the following auxiliary mean value relation holds:*

$$N^1\mathbf{u}(x) = W_1\mathbf{u}^s(x) + W_2\mathbf{u}^p(x), \tag{5.32}$$

where

$$W_1 = a\frac{\sinh(rk_s)}{rk_s} + b\left\{\frac{\cosh(rk_s)}{(rk_s)^2} - \frac{\sinh(rk_s)}{(rk_s)^3}\right\},$$

$$W_2 = (a+b)\frac{\sinh(rk_p)}{rk_p} - 2b\left\{\frac{\cosh(rk_p)}{(rk_p)^2} - \frac{\sinh(rk_p)}{(rk_p)^3}\right\},$$

$$a = 1 - b, \quad b = 3\beta, \quad \beta = \frac{5(\lambda+\mu)}{2(\lambda+4\mu)}.$$

Proof. Using the expansion (5.26) and the properties (5.29), (5.30) (in $\mathbb{R}^3$), we obtain:

$$N^1\mathbf{u}(x) - \mathbf{u}(x) = \sum_{n=1}^{\infty} \frac{r^{2n}}{(2n+1)!(2n+3)} \times \left\{(2n+3+2n\beta)k_s^{2n}\mathbf{u}^s - 3n\beta k_p^{2n}\mathbf{u}^p\right\}.$$

Summing up the series in the right-hand side of this equality yields (5.32). □

Theorem 5.7. The mean value relation.

Assume that the parameter ν *satisfies the inequalities* $k_p r < \pi$ *and* $k_s r < \pi$ *for each* $S(x,r) \subset G$. *Then the general solution to* (5.19) *in* $B(x,r) \subset \mathbb{R}^3$ *satisfies the following mean value relation*

$$\mathbf{u}(x) = \frac{(W_2 - W_1)N_r\mathbf{u} + \left[\frac{\sinh(rk_s)}{rk_s} - \frac{\sinh(rk_p)}{rk_p}\right]N^1\mathbf{u}}{\frac{\sinh(rk_s)}{rk_s}W_2 - \frac{\sinh(rk_p)}{rk_p}W_1} \tag{5.33}$$

or in a standard integral-matrix form

$$\mathbf{u}(x) = \frac{1}{4\pi}\int\limits_{\Omega} \mathbb{K}(x,s)\mathbf{u}(x+rs)d\Omega \ ,$$

where the entries of the matrix $\mathbb{K}$ *are given by*

$$k_{ij} = \left(\frac{\operatorname{sh} rk_s}{rk_s}W_2 - \frac{\operatorname{sh} rk_p}{rk_p}W_1\right)^{-1}$$

$$\times\left\{\left(W_2 - W_1 + a\left[\frac{\operatorname{sh} rk_s}{rk_s} - \frac{\operatorname{sh} rk_p}{rk_p}\right]\right)\delta_{ij} + b\left[\frac{\operatorname{sh} rk_s}{rk_s} - \frac{\operatorname{sh} rk_p}{rk_p}\right]s_i s_j\right\} . \tag{5.34}$$

Proof. Let us consider the system of equations (5.31), (5.32) and $\mathbf{u}(x) = \mathbf{u}^p(x) + \mathbf{u}^s(x)$. Solving this system for $\mathbf{u}$, we get the desired relation. □

Remark 5.1.

This statement can be extended to the general case of n dimensions.

Let ν be a complex number, the parameter of the equation

$$\Delta^*\mathbf{u}(x) + \nu\mathbf{u}(x) = 0, \quad x \in G \ ,$$

satisfying the conditions $\operatorname{Re} k_s r < \pi$ and $\operatorname{Re} k_p r < \pi$ where $k_p = \sqrt{\nu/(\lambda + 2\mu)}$ and $k_s = \sqrt{\nu/\mu}$ for any $S(x,r) \subset G$.

The spherical mean value relation under these assumptions has the form (see [29])

$$u_i(x) = \frac{1}{C\omega_n} \int\limits_{\Omega} \Big(A + Bs_is_j\Big) u_j(x + rs) d\Omega \ , \quad i = 1, 2, \ldots, n \ , \tag{5.35}$$

where

$$A = \tau_\alpha(\xi) - (n-1)\tau_{\alpha+1}(\xi) - \tau_{\alpha+1}(\eta) \ ,$$

$$B = \tau_\alpha(\eta) - \tau_\alpha(\xi) \ ,$$

and

$$C = 2^\alpha \Gamma(\alpha + 1)\{\tau_\alpha(\eta)[\tau_\alpha(\xi) - (n-1)\tau_{\alpha+1}(\xi)] - \tau_\alpha(\xi)\tau_{\alpha+1}(\eta)\} \ .$$

Here $\xi = k_p r$, $\eta = k_s r$, and $\tau_\alpha(z) = z^{-\alpha} J_\alpha(z)$.

Note that the function $\tau_\alpha(z)$ is related to $W_\alpha(z)$ through

$$W_\alpha(z) = \Gamma(\alpha/2) 2^{\alpha/2-1} \tau_{\alpha/2-1}(z)$$

and it is therefore not difficult to rewrite the relevant spherical mean value relation in terms of the functions w_n and w_{n+2}. Indeed, the solution to (5.19) satisfies the same relation (5.35) with

$$A = w_n(\xi) - \frac{n-1}{n} w_{n+2}(\xi) - \frac{1}{n} w_{n+2}(\eta) \ ,$$

$$B = w_n(\eta) - w_n(\xi) \ ,$$

$$C = w_n(\eta)\Big(w_n(\xi) - \frac{n-1}{n} w_{n+2}(\xi)\Big) - \frac{1}{n} w_{n+2}(\eta) w_n(\xi) \ .$$

5.3. Thermo–elastic equation

In this section we deal with a homogeneous isotropic medium G, whose state is characterized by the classical thermo–elastic equation (there are no body forces and no thermal sources):

$$\Delta \mathbf{u}(x) + \alpha \operatorname{grad} \operatorname{div} \mathbf{u}(x) - \gamma \operatorname{grad} \theta(x) = 0, \tag{5.36}$$

$$\Delta \theta(x) = 0, \quad x \in G \subset \mathbb{R}^3, \tag{5.37}$$

where $\mathbf{u} = (u_1(x), \ldots, u_3(x))$, u_i, θ are real-valued regular functions, $\gamma = (2\mu + 3\lambda)c$, c is the coefficient of linear heat expansion [21].

We obtain now a mean value relation for (5.36), (5.37) which generalizes the result (5.33) (see [43]).

Theorem 5.8.

The general solution to the system (5.36), (5.37) *satisfies the following mean value relation*

$$\mathbf{u}(x) = N^1\mathbf{u}(x) - \frac{3r\gamma}{\omega_3(3+\alpha)} \int\limits_{\Omega(x,1)} s\theta(x+rs)\, d\Omega(s)\ , \qquad (5.38)$$

$$\theta(x) = N_r\theta$$

where s *is the direction cosine vector.*

Proof. For the harmonic function $\theta(x)$, we have

$$\theta(x) = N_B\theta, \qquad (5.39)$$

where $N_B\theta$ is the volume integral of $\theta(x)$ over the ball $B(x,r)$ [4]. Thus, for the harmonic function $\partial\theta/\partial x_k$, the relation (5.39) takes the form

$$\frac{\partial}{\partial x_k}\theta(x) = \frac{3}{4\pi r^3}\int\limits_{B(x,r)} \frac{\partial\theta(y)}{\partial x_k}\, dy = \frac{3}{4\pi r}\int\limits_{\Omega(x,1)} \frac{x_k}{r}\theta(x+rs)\, d\Omega,\quad k=1,2,3,$$

that is

$$\frac{\partial}{\partial x_k}\theta(x) = \frac{3}{4\pi r}\int\limits_{\Omega(x,1)} s_k\theta(x+rs)\, d\Omega, \qquad k=1,2,3. \qquad (5.40)$$

By $\Delta^2 u_i = 0$ and from (5.21), (5.40) we obtain

$$\begin{aligned} N^1\mathbf{u}(x) - \mathbf{u}(x) &= \frac{r^2(\Delta + \alpha\,\mathrm{grad\,div})\mathbf{u}(x)}{3+\alpha} + \sum_{n=2}^{\infty}\frac{r^{2n}}{(2n+1)!(2n+3)} \\ &\times \left\{\frac{15n\gamma\Delta^{n-1}\mathrm{grad}\,\theta(x)}{3+\alpha} - 3(n-1)\Delta^n\mathbf{u}(x)\right\} \\ &= \frac{\gamma r^2\mathrm{grad}\ \theta(x)}{3+\alpha} = \frac{3\gamma r}{4\pi(3+\alpha)}\int\limits_{\Omega(x,1)} s\theta(x+rs)\, d\Omega(s), \end{aligned}$$

and (5.38) is proved. □

Chapter 6

The generalized Poisson formula for the Lamé equation

6.1. Plane elasticity

In this section we shall deal with the plane elasticity theory. Suppose a homogeneous isotropic medium $G \subset \mathbb{R}^2$ is given, for which

$$\Delta^* \mathbf{u} = 0, \tag{6.1}$$

where

$$\Delta^* \equiv \mu\Delta + (\lambda + \mu)\text{grad div}$$

is the Lamé operator; $\mathbf{u} = (u_1(x), u_2(x))$ is a vector of displacements whose components are real-valued regular functions.

Let us suppose that we have a circle $S(x_0, R)$ of radius R centered at x_0, lying in $\bar{G}$. Let x (with the polar coordinates (ρ, φ')) be any interior point of the disk $K(x_0, R)$, $(K(x_0, R) = \{x : |x_0 - x| \leq R\})$, and let y be a point situated on the circle $S(x_0, R)$, whose coordinates are (R, θ), where $\theta = \varphi' + \alpha$, and let z be defined by $z = y - x$, $Z = |z|$; ψ is the angle between x and z.
Let us introduce the notation

$$s_1 = \cos\psi, \quad s_2 = \sin\psi,$$

$$\hat{s}_1 = \cos\varphi, \quad \hat{s}_2 = \sin\varphi, \quad \varphi = \psi + \varphi',$$

where $\hat{s}_i$ and s_i $(i = 1, 2)$ are the direction cosines of the vector z,

$$k = \rho/R,$$

$$J = \sqrt{1 - k^2 \sin^2\psi}\,.$$

The next relations follow from the above definitions immediately

$$Z^2 = R^2 + \rho^2 - 2R\rho\cos\alpha, \tag{6.2}$$

hence

$$Z = -\rho\cos\psi + \sqrt{R^2 - \rho^2\sin^2\psi},$$

and

$$Z_k = Z/R = -ks_1 + J, \tag{6.3}$$

$$\begin{aligned} \cos\alpha &= Z/R\cos\psi + \rho/R = Z_k s_1 + k\ , \\ \sin\alpha &= Z/R\sin\psi = Z_k s_2, \end{aligned} \tag{6.4}$$

$$\begin{aligned} \hat{s}_1 &= (R\cos\theta - \rho\cos\varphi')/Z\ , \\ \hat{s}_2 &= (R\sin\theta - \rho\sin\varphi')/Z\ . \end{aligned} \tag{6.5}$$

To prove the main result we first represent some weighted spherical means in the form of a series of operators.

Theorem 6.1.

For any real-valued biharmonic analytical function $u(x)$ in the domain G, the following representations hold for an arbitrary $S(x_0, r) \subset G$

$$\frac{1}{\omega_2 R}\int\limits_{S(x_0,R)} \frac{\hat{s}_1^2 u(y)}{Z^4}\,dS = u(x)\frac{R^2 + 2\rho^2\cos^2\varphi'}{2(R^2-\rho^2)^3} + \frac{\partial u(x)}{\partial n}\frac{(2+\cos 2\varphi')\rho}{4(R^2-\rho^2)^2}$$

$$-\frac{\partial u(x)}{\partial t}\frac{\sin 2\varphi'}{4(R^2-\rho^2)^2} + \frac{\Delta u(x)}{16(R^2-\rho^2)} + \frac{1}{8(R^2-\rho^2)}L_1 u(x)$$

$$+\sum_{p=4}^{\infty}\frac{(-1)^p\rho^{p-4}}{2^4(p-1)!}\frac{\partial^{p-4}}{\partial n^{p-4}}L_1\Delta u(x), \tag{6.6}$$

where

$$L_1 = \frac{\partial^2}{\partial n^2}\cos^2\varphi' + \frac{\partial^2}{\partial t^2}\sin^2\varphi' - \frac{\partial^2}{\partial n\partial t}\sin 2\varphi',$$

while

$$\frac{1}{\omega_2 R}\int\limits_{S(x_0,R)} \frac{\hat{s}_1\hat{s}_2 u(y)}{Z^4}\,dS = u(x)\frac{\sin 2\varphi'}{2}$$

$$+\frac{\partial u(x)}{\partial n}\frac{\rho\sin 2\varphi'}{4(R^2-\rho^2)^2}+\frac{\partial u(x)}{\partial t}\frac{\rho\cos 2\varphi'}{4(R^2-\rho^2)^2}+\frac{1}{8(R^2-\rho^2)}L_2 u(x)$$

$$+\sum_{p=4}^{\infty}\frac{(-1)^p\rho^{p-4}}{2^4(p-1)!}\frac{\partial^{p-4}}{\partial n^{p-4}}L_2\Delta u(x), \tag{6.7}$$

where

$$L_2=\frac{1}{2}\Big(\frac{\partial^2}{\partial n^2}-\frac{\partial^2}{\partial t^2}\Big)\sin 2\varphi'+\frac{\partial^2}{\partial n\partial t}\cos 2\varphi',$$

and

$$\frac{1}{\omega_2 R}\int\limits_{S(x_0,R)} \frac{u(y)}{Z^4}\,dS$$
$$=\frac{(R^2+\rho^2)u(x)}{(R^2-\rho^2)^3}+\frac{\partial u(x)}{\partial n}\frac{\rho}{(R^2-\rho^2)^2}+\frac{\Delta u(x)}{4(R^2-\rho^2)}, \tag{6.8}$$

$$\frac{1}{\omega_2 R}\int\limits_{S(x_0,R)} \frac{s_1 u(y)}{Z^3}\,dS = \frac{\rho u(x)}{(R^2-\rho^2)^2}$$
$$+\frac{\partial u(x)}{\partial n}\frac{1}{2(R^2-\rho^2)}+\sum_{p=3}^{\infty}\frac{(-1)^{p+1}\rho^{p-3}}{2^3(p-1)!}\frac{\partial^{p-2}\Delta u(x)}{\partial n^{p-2}}, \tag{6.9}$$

$$\frac{1}{\omega_2 R}\int\limits_{S(x_0,R)} \frac{s_2 u(y)}{Z^3}\,dS$$
$$=\frac{\partial u(x)}{\partial t}\frac{1}{2(R^2-\rho^2)}+\sum_{p=3}^{\infty}\frac{(-1)^{p+1}\rho^{p-3}}{2^3(p-1)!}\frac{\partial^{p-4}}{\partial n^{p-4}}\frac{\partial^2\Delta u(x)}{\partial n\partial t}. \tag{6.10}$$

Proof. It suffices to prove any of the formulae (6.6)-(6.10), say, formula (6.6). Let us locate the origin of coordinates (x_1, x_2) at the point x_0, and introduce the coordinates (n, t) with the origin at x and turned with respect to the axes (x_1, x_2) through the angle φ'. Then the Taylor series expansion of the function u at any point $x \in K(0, R)$ is

$$u(y) = u(x + Zs) = u(x) + \frac{\partial u(x)}{\partial n} Z s_1 + \frac{\partial u(x)}{\partial t} Z s_2 + \cdots + \frac{Z^p}{p!} \sum_{\alpha=0}^{p} C_p^\alpha \frac{\partial^p u}{\partial n^\alpha \partial t^{p-\alpha}} s_1^\alpha s_2^{p-\alpha} + \cdots . \tag{6.11}$$

Let us multiply both sides of (6.11) by the weight $\hat{s}_1^2 / Z^4$ and integrate the result over the circle $S(0, R)$. Hence, due to the expansion

$$Z^p = \sum_{j=0}^{p} C_p^j R^p (-k^2)^{p-j} s_1^{p-j} J^j$$

we get

$$\frac{1}{\omega_2 R} \int\limits_{S(0,R)} \frac{\hat{s}_1^2}{Z^4} u(y)\, dS = \frac{u(x)}{\omega_2} \int\limits_{S(0,R)} \frac{\hat{s}_1^2}{Z^4}\, dS + \cdots + \frac{1}{\omega_2} \sum_{\alpha=0}^{p} C_p^\alpha \frac{\partial^p u(x)}{\partial n^\alpha \partial t^{p-\alpha}} \times \sum_{j=0}^{p-3} C_{p-3}^j \frac{R^{p-3}}{p!} (-k^2)^{p-j-3} \int\limits_{S(0,R)} s_1^{\alpha+p-j-3} s_2^{p-\alpha} J^j \frac{\hat{s}_1^2}{Z}\, dS + \cdots . \tag{6.12}$$

All transformations of the series (6.11) (integration term-to-term, summation order change) are possible by virtue of the hypotheses of the theorem.

The Jacobian of the transformation of a circle element dS to an angle element is equal to Z/J. Since the function J is positive and symmetric relative to the point π in the interval $[0, 2\pi]$, the integrals in (6.12) differ from zero if the powers of s_i, $(i = 1, 2)$ are even. Consequently, in (6.12), only J^{2j} $(j = 1, 2, \cdots)$ should be considered. Hence J^{2j} can be decomposed using the binomial formula. To make the presentation clear we treat separately the sums involving the even and odd terms. The even terms of the series in (6.12) are (for $p > 3$)

$$\sum_{p=3}^{\infty} \frac{R^{2p-3}}{\omega_2(2p)!} \sum_j C_{2p-3}^{2j-1} k^{2p-2j-4} \sum_{l=0}^{j} C_j^l (-k^2)^{j-l} \Big\{ \sum_{\alpha=0}^{p} C_{2p}^{2\alpha} \frac{\partial^{2p} u}{\partial n^{2\alpha} \partial t^{2p-2\alpha}}$$
$$\times \int_0^{2\pi} (s_1^2 \cos^2\varphi' + s_2^2 \sin^2\varphi') s_1^{2p-2j+2\alpha-4} s_2^{2p-2\alpha+2j-2l} \, d\psi$$
$$- \sum_{\alpha=0}^{p-1} C_{2p}^{2\alpha+1} \frac{\partial^{2p} u}{\partial n^{2\alpha+1} \partial t^{2p-2\alpha-1}} \sin 2\varphi' \int_0^{2\pi} s_1^{2p-2j+2\alpha-2} s_2^{2p-2\alpha+2j-2l} \, d\psi \Big\}. \tag{6.13}$$

The remaining integrals can be taken from tables and are expressed in terms of Euler integrals,

$$\frac{1}{\omega_2} \int_0^{2\pi} s_1^{2n} s_2^{2m} \, d\psi = \frac{(2n-1)!!(2m-1)!!}{(2n+2m)!!} .$$

Let us roll up the sum over α in (6.13). To this end we introduce the independent parameters β and γ such that

$$\frac{(2n-2m-1)!!}{(2m-1)!!}(-2)^{-n} = \frac{\partial^n}{\partial \beta^n} \left(\beta^{-(2m+1)/2} \right) \Big|_{\beta=1} .$$

The same property holds for the parameter γ. Then (6.13) takes the form

$$\sum_{p=3}^{\infty} \frac{R^{2p-3}}{(2p)!} \sum_j C_{2p-3}^{2j-1} k^{2p-2j-4} \sum_{l=0}^{j} C_j^l (-k^2)^{j-l} \frac{(2p-1)!!}{(4p-2l-2)!!} (-2)^{p-1-l}$$
$$\times \Big\{ \sum_{\alpha=0}^{p} C_p^\alpha \frac{\partial^{2p} u}{\partial n^{2\alpha} \partial t^{2p-2\alpha}} (A \cos^2\varphi' + B \sin^2\varphi')$$
$$- \sum_{\alpha=0}^{p-1} C_{p-1}^\alpha \frac{\partial^{2p} u}{\partial n^{2\alpha+1} \partial t^{2p-2\alpha-1}} 2pC \sin 2\varphi' \Big\},$$

where

$$A = \left(\beta^{-(2\alpha+1)/2} \right)_\beta^{(p-j-1)} \left(\gamma^{-(2p-2\alpha+1)/2} \right)_\gamma^{(j-l)} \Big|_{\gamma=\beta=1},$$
$$B = \left(\beta^{-(2\alpha+1)/2} \right)_\beta^{(p-j-2)} \left(\gamma^{-(2p-2\alpha+1)/2} \right)_\gamma^{(j-l+1)} \Big|_{\gamma=\beta=1},$$

$$C = \left(\beta^{-(2\alpha+3)/2}\right)_{\beta}^{(p-j-2)} \left(\gamma^{-(2p-2\alpha+1)/2}\right)_{\gamma}^{(j-l)}\Big|_{\gamma=\beta=1}.$$

Changing the order of differentiation with respect to the parameters and the summation over α, we can then rewrite the sum over α as

$$\sum_{p=3}^{\infty} \frac{R^{2p-3}}{(2p)!} \sum_j C_{2p-3}^{2j-1} k^{2p-2j-4} \sum_{l=0}^{j} C_j^l (-k^2)^{j-l} \frac{(2p-1)!!}{(4p-2l-2)!!} (-2)^{p-1-l}$$
$$\times \Big(\frac{\partial^{p-j-1}}{\partial \beta^{p-j-1}} \frac{\partial^{j-l}}{\partial \gamma^{j-l}} \big(\frac{1}{\sqrt{\beta\gamma}} L^p u\big) \cos^2 \varphi'$$
$$+ \frac{\partial^{p-j-2}}{\partial \beta^{p-j-2}} \frac{\partial^{j-l+1}}{\partial \gamma^{j-l+1}} \big(\frac{1}{\sqrt{\beta\gamma}} L^p u\big) \sin^2 \varphi'$$
$$-p \frac{\partial^{p-j-2}}{\partial \beta^{p-j-2}} \frac{\partial^{j-l}}{\partial \gamma^{j-l}} \big(\frac{1}{(\beta\gamma)^{3/2}} L^{p-1} \frac{\partial^2 u}{\partial n \partial t}\big) \sin 2\varphi' \Big)\Big|_{\beta=\gamma=1},$$

where

$$L = \frac{1}{\beta}\frac{\partial^2}{\partial n^2} + \frac{1}{\gamma}\frac{\partial^2}{\partial t^2}.$$

It is easy to see that L is the Laplace operator if $\beta = \gamma = 1$. Hence, if after the differentiation with respect to β and γ, the operator L acts on the function u more then twice, then this term vanishes. Therefore, only the terms with $l = 0$ do not vanish. Among these, the non-zero terms appear only when the parametric derivatives of L are taken, since otherwise there are terms involving L^q with $q > 1$ (recall that $\Delta^2 u = 0$). Now,

$$\frac{\partial^{n-m-1}}{\partial \beta^{n-m-1}} \frac{\partial^m}{\partial \gamma^m} \Big(\frac{L^n}{\beta^l \gamma^k}\Big)\Big|_{\beta=\gamma=1} = n!(-1)^{n-1} \frac{\partial^{2n-2m-2}}{\partial n^{2n-2m-2}} \frac{\partial^{2m}}{\partial t^{2m}},$$

$$\frac{\partial^{2m}}{\partial t^{2m}} \Delta u = \Big(\Delta - \frac{\partial^2}{\partial n^2}\Big)^m \Delta u = \sum_{l=0}^{m} C_m^l (-1)^l \frac{\partial^{2l}}{\partial n^{2l}} \Delta^{m-l+1} u$$
$$= (-1)^m \frac{\partial^{2m} \Delta u}{\partial n^{2m}}.$$

These relations are proved by direct differentiation.
The property

$$\sum_{j=0}^{[(p-1)/2]} C_p^{2j+1} = 2^{p-1}$$

was used above.

In order to get the desired result (6.6) it remains to add the odd terms transformed by the same procedure and the first terms with $p < 3$ to the expression for even p. The first term values are given by the next formulae, which are proved using a simple technique:

$$
\begin{aligned}
&\frac{1}{2\pi R}\int\limits_{S(0,R)} \frac{1}{Z^4}\,dS = \frac{R^2+\rho^2}{(R^2-\rho^2)^3}\,,\\
&\frac{1}{2\pi R}\int\limits_{S(0,R)} \frac{s_1}{Z^3}\,dS = \frac{\rho}{(R^2-\rho^2)^2}\,,\\
&\frac{1}{2\pi R}\int\limits_{S(0,R)} \frac{s_i^2}{Z^2}\,dS = \frac{1}{2(R^2-\rho^2)}, \quad i=1,2\,,\\
&\frac{1}{2\pi R}\int\limits_{S(0,R)} \frac{s_1^2}{Z^4}\,dS = \frac{R^2+2\rho^2}{2(R^2-\rho^2)^3}\,,\\
&\frac{1}{2\pi R}\int\limits_{S(0,R)} \frac{s_2^2}{Z^4}\,dS = \frac{R^2}{2(R^2-\rho^2)^3}\,,\\
&\frac{1}{2\pi R}\int\limits_{S(0,R)} \frac{s_1^3}{Z^3}\,dS = \frac{3\rho}{4(R^2-\rho^2)^2}\,,\\
&\frac{1}{2\pi R}\int\limits_{S(0,R)} \frac{s_1 s_2^2}{Z^3}\,dS = \frac{\rho}{4(R^2-\rho^2)^2}\,,\\
&\frac{1}{2\pi R}\int\limits_{S(0,R)} \frac{s_i^2 s_j^2}{Z^2}\,dS = \frac{1+2\delta_{ij}}{8(R^2-\rho^2)}, \quad i=1,2, \quad j=1,2.
\end{aligned}
\tag{6.14}
$$

Thus the required result (6.6) is deduced. The formulae (6.7)-(6.10) are proved in a similar way. □

The following auxiliary result is used in Theorem 6.2.

Lemma 6.1.

For the solution of the Lamé equation $\mathbf{u}$ *the properties*

$$\Delta\Big(\frac{\partial u_1}{\partial n}+\frac{\partial u_2}{\partial t}\Big)=0,$$

$$\Delta\Big(\frac{\partial u_1}{\partial t}-\frac{\partial u_2}{\partial n}\Big)=0$$

hold.

Proof. It is known [21] that

$$\Delta\Big(\frac{\partial u_1}{\partial x_2}-\frac{\partial u_2}{\partial x_1}\Big)=0,$$

$$\Delta\Big(\frac{\partial u_1}{\partial x_1}+\frac{\partial u_2}{\partial x_2}\Big)=0.$$

Rewriting these equations in the coordinates (n,t) and taking the corresponding linear combinations we get the result. □

Theorem 6.2.

For any analytical in the domain G solution of the equation (6.1) *and for every circle $K(x_0,R)$, lying in $\bar{G}$ the following auxiliary mean value relation holds at an arbitrary point x, $x\in K(x_0,R)$*

$$\frac{1}{\omega_2 R}\int\limits_{S(x_0,R)}\Big(\hat{s}_i\hat{s}_j-\frac{\delta_{ij}}{2}\Big)\frac{u_j(y)}{Z^4}\,dS-\frac{\rho}{2(R^2-\rho^2)\omega_2 R}\int\limits_{S(x_0,R)}\frac{a_{ij}u_j(y)}{Z^3}\,dS$$
$$=-\Delta u_i(x)\frac{1+2/\alpha}{16(R^2-\rho^2)},\quad i,j=1,2, \tag{6.15}$$

where

$$A=\Big\{a_{ij}\Big\}_{i,j=1,2}=\begin{pmatrix}\cos(2\varphi'+\psi) & \sin(2\varphi'+\psi)\\ \sin(2\varphi'+\psi) & -\cos(2\varphi'+\psi)\end{pmatrix}.$$

Proof. Let us consider the integral

$$I=\frac{1}{\omega_2}\int\limits_{S(x_0,R)}\Big(\hat{s}_1\hat{s}_j-\frac{\delta_{ij}}{2}\Big)\frac{u_j(y)}{Z^4}\,dS.$$

By Theorem 6.1, this integral can be expressed as a series of differential operators. Since the function u_i is biharmonic and the first equation of the Lamé system in the coordinates (n,t) is

$$\Delta u_1 + \alpha\Big(\cos^2\varphi'\frac{\partial^2}{\partial n^2} + \sin^2\varphi'\frac{\partial^2}{\partial t^2} - \sin 2\varphi'\frac{\partial^2}{\partial n\partial t}\Big)u_1$$
$$+ \alpha\Big(\frac{1}{2}\sin 2\varphi'\big(\frac{\partial^2}{\partial n^2} - \frac{\partial^2}{\partial t^2}\big) + \cos 2\varphi'\frac{\partial^2}{\partial n\partial t}\Big)u_2 = 0,$$

then the integral I is transformed to

$$I = \frac{\rho^2}{2(R^2-\rho^2)^3}\Big(\cos 2\varphi' u_1 + \sin 2\varphi' u_2\Big) - \Delta u_1\frac{1+2/\alpha}{16(R^2-\rho^2)}$$
$$+\frac{\rho\cos 2\varphi'}{4(R^2-\rho^2)^2}\Big\{\frac{\partial u_1}{\partial n} + \frac{\partial u_2}{\partial t}\Big\} + \frac{\rho\sin 2\varphi'}{4(R^2-\rho^2)^2}\Big\{\frac{\partial u_2}{\partial n} - \frac{\partial u_1}{\partial t}\Big\}. \tag{6.16}$$

The same transformations are made with the integrals

$$\frac{1}{\omega_2}\int\limits_{S(x_0,R)}\frac{s_1u_1+s_2u_2}{Z^3}\,dS, \quad \frac{1}{\omega_2}\int\limits_{S(x_0,R)}\frac{s_1u_2-s_2u_1}{Z^3}\,dS.$$

On the basis of Theorem 6.1, these integrals are expanded and then Lemma 6.1 is applied:

$$\frac{1}{\omega_2}\int\limits_{S(x_0,R)}\frac{s_1u_1+s_2u_2}{Z^3}\,dS = \frac{\rho}{(R^2-\rho^2)^2}u_1 + \frac{1}{2(R^2-\rho^2)}\Big\{\frac{\partial u_1}{\partial n} + \frac{\partial u_2}{\partial t}\Big\},$$
$$\frac{1}{\omega_2}\int\limits_{S(x_0,R)}\frac{s_1u_2-s_2u_1}{Z^3}\,dS = \frac{\rho}{(R^2-\rho^2)^2}u_2 + \frac{1}{2(R^2-\rho^2)}\Big\{\frac{\partial u_2}{\partial n} - \frac{\partial u_1}{\partial t}\Big\}. \tag{6.17}$$

Extracting the expressions in the braces in (6.17) and substituting them into (6.16) we obtain the first relation ($i = 1$) of the system (6.15). The second relation ($i = 2$) is proved analogously. □

Theorem 6.3.

For any analytical in G solution $\mathbf{u}(x)$ of the Lamé equation and for any circle $S(x_0, R) \subset G$ at an arbitrary interior point of $K(x_0, R)$, the following mean value relations hold:

$$
\begin{aligned}
u_i(x) = \frac{1-k^2}{2\pi} \sum_{j=1}^{2} \int_0^{2\pi} \left(\delta_{ij} + \frac{2}{\sigma'} \hat{s}_i \hat{s}_j\right) \frac{u_j(\theta)}{Z_k^2}\, d\theta \\
+ \frac{1-k^2}{2\pi\sigma'} \sum_{j=1}^{2} \int_0^{2\pi} \frac{b_{ij} u_j(\theta)}{Z_k^2}\, d\theta \\
i = 1, 2,
\end{aligned}
\tag{6.18}
$$

where

$$
\left\{b_{ij}\right\} = \begin{pmatrix} 2\cos^2\theta - k\cos(\theta+\varphi') & \sin 2\theta + k\sin(\theta+\varphi') \\ \sin 2\theta + k\sin(\theta+\varphi') & 2\sin^2\theta - k\cos(\theta+\varphi') \end{pmatrix},
$$

$\hat{s}_i$, Z_k *are defined in the formulae* (6.2),(6.3), *and*

$$
\sigma' = 1 + \frac{2\mu}{\lambda+\mu}.
$$

Proof. Let $i = 1$, (the case $i = 2$ is deduced in a similar manner). We introduce the notation

$$
u_i(y) = g_i(y), \quad \Delta u_i(y) = g_i^1(y), \quad i = 1, 2, \quad y \in S(x_0, R).
$$

It is known [55] that any biharmonic function can be represented through two harmonic functions

$$
u_i(x) = (R^2 - \rho^2) v_i(x) + w_i(x), \tag{6.19}
$$

where the functions v_i and w_i are the solutions to the following problems

$$
\begin{cases} \Delta w_i(x) = 0, & x \in K(x_0, R) \\ w_i(y) = g_i(y), & y \in S(x_0, R) \end{cases}, \tag{6.20}
$$

and

$$
\begin{cases} \Delta v_i(x) = 0, & x \in K(x_0, R) \\ \left. \dfrac{\partial v_i(y)}{\partial r} + \dfrac{1}{r} v_i \right|_{r=R} = -\dfrac{g_i^1(y)}{4R}, & y \in S(x_0, R) \end{cases}. \tag{6.21}
$$

The Poisson formula yields the solution to the problem (6.20):

$$w_i = \frac{(1-k^2)R}{2\pi} \int\limits_{S(x_0,R)} \frac{g_i(y)}{Z^2}\, dS. \tag{6.22}$$

As to the problem (6.21) – we treat it with the help of the separation of variables [33].

$$v_i(\rho,\varphi') = \frac{\rho}{2}A_0^i + \sum_{n=1}^{\infty} \frac{\rho^n}{R^{n-1}(n+1)}(A_n^i \cos n\varphi' + B_n^i \sin\varphi'),$$

$$A_n^i = -\frac{1}{4\pi R}\int\limits_0^{2\pi} g_i^1(y)\cos n\theta\, d\theta,$$

$$B_n^i = -\frac{1}{4\pi R}\int\limits_0^{2\pi} g_i^1(y)\sin n\theta\, d\theta,$$

$$A_0^i = -\frac{1}{4\pi R}\int\limits_0^{2\pi} g_i^1(y)\, d\theta.$$

Then

$$v_i(\rho,\varphi') = -\frac{1}{4\pi}\int\limits_0^{2\pi} g_i^1(y)\sum_{n=1}^{\infty}\left(\frac{k^n}{n+1}\cos n(\theta-\varphi') + \frac{1}{2}\right) d\theta.$$

The change of integration and summation order is possible since for $k < 1$ the series is uniformly convergent with respect to $\alpha = \theta - \varphi$. It is not difficult to calculate its sum [15]

$$v_i(\rho,\varphi') = -\frac{1}{16\pi k}\int\limits_0^{2\pi} g_i^1(y)\int\limits_0^{k}\frac{1-t^2}{1-2t\cos\alpha+t^2}\, dt\, d\theta.$$

In the region $[0;2\pi]\times[0;k]$ the integrated functions are continuous, so we can change the order of integration:

$$v_i(\rho,\varphi') = -\frac{1}{16\pi k}\int\limits_0^{k}\left[(1-t^2)\int\limits_0^{2\pi}\frac{g_i^1(y)}{1-2t\cos\alpha+t^2}\, d\theta\right] dt.$$

Now in the square brackets is the Poisson integral formula for the harmonic function Δu_i (indeed, $\Delta(\Delta u_i) = 0$); therefore

$$v_i(\rho, \varphi') = -\frac{1}{8kR} \int_0^\rho \Delta u_i(t_1, \varphi')\, dt_1 \, .$$

But for any point $x \in K(x_0, R)$, $x = (t_1, \varphi')$, the value of $\Delta u_i(x)$ is given by Theorem 6.2. Hence, we can write for $i = 1$:

$$v_1(\rho, \varphi') = \frac{2}{\pi k \sigma' R^2} \int_0^k (1 - t^2) \int_0^{2\pi} \Big(\frac{(\hat{s}_1^2 - 0.5)}{Z_t^4} u_1 + \frac{\hat{s}_1 \hat{s}_2}{Z_t^4} u_2 \Big)\, d\theta\, dt$$

$$-\frac{1}{\pi k \sigma' R^2} \int_0^k t \int_0^{2\pi} \frac{1}{Z_t^4} \Big(\cos(\theta + \varphi')\, u_1 + \sin(\theta + \varphi')\, u_2 \Big)\, d\theta\, dt$$

$$+\frac{1}{\pi k \sigma' R^2} \int_0^k t^2 \int_0^{2\pi} \frac{1}{Z_t^4} \Big(\cos 2\varphi'\, u_1 + \sin 2\varphi'\, u_2 \Big)\, d\theta\, dt,$$

where

$$Z_t^2 = 1 - 2t \cos\alpha + t^2.$$

Changing the order of integration we can then calculate the integrals over t on the basis of tables [15]. Adding the expression obtained for $v_i(x)$ with the weight $(1 - k^2)$ and for $w_i(x)$ from (6.22), we get the desired relation. □

Corollary 6.1.

If $k = 0$ (i.e., $x = x_0$ is the center of the circle $K(x_0, R)$), then the relations (6.19) become the mean value relations for the Lamé equation.

6.2. Generalized spatial Poisson formula for the Lamé equation

We turn now to the $3D$ case (see [52]). Suppose for simplicity that we have a ball $B(0, R)$ of radius R centered at the origin of coordinates. Let M be an arbitrary point with the spherical coordinates (ρ, θ, φ) inside the ball, and let P be a point on the sphere $S = \partial B(0, R)$.

We denote by W the distance $|M-P|$ and let $\hat{s}=(\hat{s}_1,\hat{s}_2,\hat{s}_3)$ be the direction cosines of the vector $P-M$. We need also a triple of coordinate axes $(\mathbf{n}_1,\mathbf{n}_2,\mathbf{n}_3)$ centered at the point M and oriented with respect to the axes $(\mathbf{x}_1,\mathbf{x}_2,\mathbf{x}_3)$ so that $\mathbf{n}_1$ coincides with the vector $M-O$.

In these new coordinates, the direction cosines of the vector $P-M$ are $s=(s_1,s_2,s_3)$ and those of the vector $P-O$ are $\alpha=(\alpha_1,\alpha_2,\alpha_3)$.

From geometrical considerations, we write down the relations

$$W^2 = R^2+\rho^2-2R\rho\alpha_1\,, \tag{6.23}$$

$$R^2 = W^2+\rho^2+2W\rho s_1\,; \tag{6.24}$$

hence

$$W = R(-ks_1+J_k), \tag{6.25}$$

where

$$J_k=\sqrt{1-k^2+k^2s_1^2},\qquad k=\rho/R.$$

The vectors s and $\hat{s}$ are related by

$$s=A^T\hat{s}, \tag{6.26}$$

where

$$A=\begin{pmatrix} \sin\theta\cos\varphi & \cos\theta\cos\varphi & -\sin\varphi\\ \sin\theta\sin\varphi & \cos\theta\sin\varphi & \cos\varphi\\ \cos\theta & -\sin\theta & 0 \end{pmatrix}, \tag{6.27}$$

and

$$s_1=W^{-1}(R\alpha_1-\rho),\quad s_2=W^{-1}R\alpha_2,\quad s_3=W^{-1}R\alpha_3. \tag{6.28}$$

To prove the main theorem we need the integrals of the solution over the averaging measure represented as a series of operators. We first prove

Theorem 6.4.

For any analytic function u which is biharmonic in a ball the following expansions hold.

First,

$$\frac{1}{4\pi R^2}\int\limits_S \frac{s_i u(P)}{W^4}\,dS = u(M)\frac{2\rho\delta_{i1}}{3R(R^2-\rho^2)^2} + \frac{\partial u(M)}{\partial n_i}\frac{1}{3R(R^2-\rho^2)}$$

$$+\sum_{p=3}^{\infty}\frac{(-1)^{p-3}}{R}\frac{(2\rho)^{p-3}}{2(2p-1)!!}\frac{\partial}{\partial n_i}\frac{\partial^{p-3}\Delta u(M)}{\partial n_1^{p-3}}, \tag{6.29}$$

next

$$\frac{1}{4\pi R^2}\int\limits_S \frac{s_i^2 u(P)}{W^5}\,dS$$

$$= u(M)\frac{5R^2-(1-8\delta_{i1})\rho^2}{15R(R^2-\rho^2)^3} + \frac{\partial u(M)}{\partial n_1}\frac{2\rho+4\rho\delta_{i1}}{15R(R^2-\rho^2)^2}$$

$$+\frac{\Delta u(M)}{30R(R^2-\rho^2)} + \sum_{p=4}^{\infty}\frac{(-1)^p}{R}\frac{(2\rho)^{p-4}}{2(2p-1)!!}\frac{\partial^2}{\partial n_i^2}\frac{\partial^{p-4}\Delta u(M)}{\partial n_1^{p-4}}, \tag{6.30}$$

and

$$\frac{1}{4\pi R^2}\int\limits_S \frac{s_i s_j u(P)}{W^5}\,dS$$

$$= \frac{\partial u(M)}{\partial n_i}\frac{2\rho(1-\delta_{i2})}{15R(R^2-\rho^2)^2} + \frac{\partial^2 u(M)}{\partial n_i \partial n_j}\frac{1}{15R(R^2-\rho^2)} \tag{6.31}$$

$$+\sum_{p=4}^{\infty}\frac{(-1)^p}{R}\frac{(2\rho)^{p-4}}{2(2p-1)!!}\frac{\partial^2}{\partial n_i \partial n_j}\frac{\partial^{p-4}\Delta u(M)}{\partial n_1^{p-4}}.$$

In the above formulae, $i = 1, 2, 3$, and $j > i$. Finally,

$$\frac{1}{4\pi R^2}\int\limits_S \frac{u(P)}{W^5}\,dS$$
$$= u(M)\frac{3R^2+\rho^2}{3R(R^2-\rho^2)^3} + \frac{\partial u(M)}{\partial n_1}\frac{2\rho}{2R(R^2-\rho^2)^2} + \frac{\Delta u(M)}{6R(R^2-\rho^2)}, \tag{6.32}$$

where δ_{ij} is Kronecker's symbol.

Proof. It suffices to prove any of the formulae (6.29) – (6.32), say, formula (6.30). Let us expand the function $u(P)$ by its Taylor series expansion in the coordinates $(\mathbf{n}_1, \mathbf{n}_2, \mathbf{n}_3)$:

$$u(P) = u(M+Ws) = u(M) + \sum_{i=1}^{3}\frac{\partial u(M)}{\partial n_i}s_i W + \dots$$
$$+\frac{W^p}{p!}\sum_{c_1+c_2+c_3=p}(c_1,c_2,c_3)\frac{\partial^p u(M)}{\partial n_1^{c_1}\partial n_2^{c_2}\partial n_3^{c_3}}s_1^{c_1}s_2^{c_2}s_3^{c_3} + \dots \tag{6.33}$$

where
$$(c_1,c_2,c_3) = \frac{(c_1+c_2+c_3)!}{c_1!c_2!c_3!}\ .$$

Let us multiply both sides of (6.33) by the weight s_1^2/W^5 and integrate them over the sphere $S(0,R)$. To this end, we pass on to integration with respect to the spherical coordinates (W,θ',φ') at the point M. Note that the vector s can be represented as

$$s_1 = \cos\varphi'\sin\theta'\ ,\ s_2 = \sin\varphi'\sin\theta'\ ,\ s_3 = \cos\theta'\ .$$

The Jacobian of the transition is given by the following lemma.

Lemma 6.2.

A surface element dS at a point P of the sphere S is given by

$$dS = \frac{W^2\sin\theta'}{J_k}\,d\varphi'\,d\theta'.$$

The proof follows from the known formulae for the change of variables [19].

Hence, (6.33) can be written as

$$\frac{1}{4\pi R^2}\int\limits_S \frac{s_i^2 u(P)}{W^5}\,dS$$
$$= I_0 + \sum_{p=4}^{\infty}\frac{1}{p!}\sum_{c_1+c_2+c_3=p}(c_1,c_2,c_3)\frac{\partial^p u(M)}{\partial n_1^{c_1}\partial n_2^{c_2}\partial n_3^{c_3}}$$
$$\times\sum_{j=0}^{p-3} C_{p-3}^j R^{p-3}(-k)^{p-3-j}\int\limits_0^{\pi}\int\limits_0^{2\pi} s_1^{p-1-j+c_1}s_2^{c_2}s_3^{c_3}J_k^{j-1}\sin\theta'\,d\theta'\,d\varphi'. \tag{6.34}$$

Here we used the expansion

$$W^p = R^p(-ks_1+J_k)^p = \sum_{j=0}^{p} R^p C_p^j(-k)^{p-j}s_1^{p-j}J_k^j.$$

The symbol I_0 stands for the sum of first terms with $p \le 3$.

Now it is convenient to partition the series over p into sums over even and odd terms and consider each case separately. We start with the case $p = 2p'$.

It is easy to verify that the integrals of the form

$$\int\limits_0^{\pi}\int\limits_0^{2\pi} s_1^{p-j+c_1-1}s_2^{c_2}s_3^{c_3}J_k^{j-1}\sin\theta'\,d\theta'\,d\varphi'$$

differ from zero only if all the powers of the directional cosines s_i are even. Hence if $p = 2p'$, then for the integrals different from zero we find

$$c_i = 2c_i', \quad i = 1,2,3, \quad j = 2j'+1$$

and we have an even power of J_k, which can then be multiplied out. Note that

$$J_k^{2j'} = (1-k^2+k^2s_1^2)^{j'} = \sum_{l=0}^{j'} C_{j'}^l(1-k^2)^l(k^2)^{j'-l}s_1^{2(j'-l)}\ .$$

Therefore, we can integrate the remaining cosines s_i by the formula [15]

$$\frac{1}{4\pi}\int_0^{\pi}\int_0^{2\pi} s_1^{2m_1}s_2^{2m_2}s_3^{2m_3}\sin\theta'\,d\theta'\,d\varphi' = \frac{(2m_1-1)!!(2m_2-1)!!(2m_3-1)!!}{(2m_1+2m_2+2m_3+1)!!}\ .$$

Let us consider the inner sum in (6.34), i.e., the sum $\sum\limits_{c_1+c_2+c_3=p}$. This sum transformed as above, takes the form

$$\frac{(2p'-1)!!}{(4p'-2l-1)!!} \times \sum_{c_1'+c_2'+c_3'=p'} (c_1', c_2', c_3') \frac{\partial^{2p'} u}{\partial n_1^{2c_1'}\partial n_2^{2c_2'}\partial n_3^{2c_3'}} \frac{(2p'-2l+2c_1'-3)!!}{(2c_1'-1)!!}\ . \tag{6.35}$$

Let us introduce in (6.35) a parameter β such that

$$\frac{(2p'-2l+2c_1'-3)!!}{(2c_1'-1)!!} = (-2)^{p'-l-1}\frac{\partial^{p'-l-1}}{\partial\beta^{p'-l-1}}\left\{\frac{1}{\beta^{(2c_1'+1)/2}}\right\}_{\beta=1}\ .$$

Placing the differentiation with respect to the parameter β outside the sign of summation over c_i', we can then contract the sum in (6.35) to get

$$\frac{(2p'-1)!!}{(2p'-2l-1)!!}\frac{\partial^{p'-l-1}}{\partial\beta^{p'-l-1}} \times\left\{\frac{1}{\sqrt{\beta}}\left(\frac{1}{\beta}\frac{\partial^2}{\partial n_1^2}+\frac{\partial^2}{\partial n_2^2}+\frac{\partial^2}{\partial n_3^2}\right)^{p'} u\right\}(-2)^{p'-l-1}\Bigg|_{\beta=1}\ . \tag{6.36}$$

Now let us consider the differentiation of the expression in braces with respect to β. In both cases we may change the order of differentiation, because the sum c_i' is finite and β is an independent parameter. Since u is a biharmonic function, i.e.

$$\left(\frac{\partial^2}{\partial n_1^2}+\frac{\partial^2}{\partial n_2^2}+\frac{\partial^2}{\partial n_3^2}\right)^2 u = 0\ ,$$

then at the point $\beta = 1$, all the terms in (6.36) with $l \neq 0$ vanish and only terms in which we take derivatives of the parenthetical factor in (6.36) will not vanish in the remaining sum. As this happens, its power reduces.

As a result, we obtain in (6.36) the following terms:

$$\frac{(2p'-1)!!}{(4p'-1)!!}2^{p'-1}p'!\frac{\partial^{2p'-2}}{\partial n_1^{2p'-2}}\Delta u.$$

It remains only to add terms transformed by the same procedure and the nonzero first terms with $p < 3$ to the expression for even p in order to get the desired formula. The first term can be integrated directly. The corresponding results are given as the following lemma.

Lemma 6.3.

The following table of integrals is true:

$$\frac{1}{4\pi R^2}\int\limits_S \frac{1}{W^5}\,dS = \frac{3R^2+\rho^2}{3R(R^2-\rho^2)^3}\,,$$

$$\frac{1}{4\pi R^2}\int\limits_S \frac{s_1}{W^4}\,dS = \frac{2\rho}{3R(R^2-\rho^2)^2}\,,$$

$$\frac{1}{4\pi R^2}\int\limits_S \frac{s_i^2}{W^3}\,dS = \frac{1}{3R(R^2-\rho^2)}\,, \quad i = 1,2,3\,,$$

$$\frac{1}{4\pi R^2}\int\limits_S \frac{s_1^2}{W^5}\,dS = \frac{5R^2+7\rho^2}{15R(R^2-\rho^2)^3}\,,$$

$$\frac{1}{4\pi R^2}\int\limits_S \frac{s_j^2}{W^5}\,dS = \frac{5R^2-\rho^2}{15R(R^2-\rho^2)^3}\,, \quad j = 2,3\,,$$

$$\frac{1}{4\pi R^2}\int\limits_S \frac{1}{W^7}\,dS = \frac{5R^4+10R^2\rho^2+\rho^4}{5R(R^2-\rho^2)^5}\,, \tag{6.37}$$

$$\frac{1}{4\pi R^2}\int\limits_S \frac{s_1^3}{W^4}\,dS = \frac{2\rho}{5R(R^2-\rho^2)^2}\,,$$

$$\frac{1}{4\pi R^2}\int\limits_S \frac{s_1 s_j^2}{W^4}\,dS = \frac{2\rho}{15R(R^2-\rho^2)^2}\,, \quad j = 2,3\,,$$

$$\frac{1}{4\pi R^2}\int\limits_S \frac{s_i^4}{W^3}\,dS = \frac{1}{5R(R^2-\rho^2)}\,, \quad i = 1,2,3\,,$$

$$\frac{1}{4\pi R^2}\int\limits_S \frac{s_i^2 s_j^2}{W^3}\,dS = \frac{1}{15R(R^2-\rho^2)}, \quad i=1,2, \quad j=2,3, \quad i<j\,,$$

where s_i, W, *and* ρ *are defined in* (6.23)-(6.28).

Proof. Let us prove the first of the formulae (6.37). The proof of each subsequent integral is based on the preceding one and on (6.23)-(6.28). Let us represent the surface element of the unit sphere as

$$d\Omega = J dW_k d\theta.$$

Taking into account that

$$W_k = \sqrt{1+k^2-2k\alpha_1} = \sqrt{1+k^2-2k\cos\varphi\sin\theta}$$

we have

$$J = \frac{W_k}{k\sin\varphi} = \frac{W_k\alpha_3}{k\alpha_2}\,.$$

Then

$$\int\limits_\Omega \frac{d\Omega}{W_k^5} = 2\int\limits_{1+k}^{1-k} \frac{1}{kW_k^4}\,dW_k$$

$$\times \int \left\{1-\cos^2\theta - \left(\frac{1+k^2-W_k^2}{2k}\right)^2\right\}^{(-1/2)} \sin\theta\; d\theta.$$

Let us consider the inner integral. It is known (see, for example, [3]) that

$$\int \left\{1-\cos^2\theta - \left(\frac{1+k^2-W_k^2}{2k}\right)^2\right\}^{(-1/2)} \sin\theta\; d\theta$$

$$= \arcsin\left(\left\{1-\left(\frac{1+k^2-W_k^2}{2k}\right)^2\right\}^{(-1/2)} \cos\theta\right).$$

Because the plane (M,O,P) coincides with $(n_1,0,n_3)$, we have at the ends of the intervals of integration,

$$\cos\theta = \sqrt{1-\left(\frac{1+k^2-W_k^2}{2k}\right)^2}, \quad -\cos\theta = \sqrt{1-\left(\frac{1+k^2-W_k^2}{2k}\right)^2}.$$

Hence,

$$\int \left\{1 - \cos^2\theta - \left(\frac{1 + k^2 - W_k^2}{2k}\right)^2\right\}^{(-1/2)} \sin\theta \, d\theta = 2 \arcsin 1 = \pi$$

and the desired integral is

$$2\pi \int_{1+k}^{1-k} \frac{1}{kW_k^4} \, dW_k = 2\pi \left[\frac{1}{3kW_k^3}\right]_{1+k}^{1-k} = 4\pi \frac{3 + k^2}{3(1 - k^2)^3}$$

which is the required result. □

Now, we return to the solution to the Lamé equation (6.1) in the domain $G \subset \mathbb{R}^3$

$$\Delta^* \mathbf{u} = 0,$$

where $\mathbf{u} = (u_1, u_2, u_3)$ is a real-valued regular vector function with the properties [21]

$$\Delta^2 u_i = 0, \quad i = 1, 2, 3, \tag{6.38}$$

$$\Delta \operatorname{div} \mathbf{u} = 0, \tag{6.39}$$

$$\Delta \operatorname{rot} \mathbf{u} = 0. \tag{6.40}$$

Theorem 6.5.

The following auxiliary mean value relation holds for the solution $\mathbf{u}(M)$ of the Lamé equations:

$$\Delta u_j(\rho, \theta, \varphi) + \frac{2u_j(\rho, \theta, \varphi)}{(R^2 - \rho^2)\sigma'}$$
$$= -\frac{30(R^2 - \rho^2)R}{4\pi R^2 \sigma'} \sum_{i=1}^{3} \int_S \frac{(\hat{s}_i \hat{s}_j - 2/5\delta_{ij})u_i(P)}{W^5} \, dS(P)$$
$$+ \frac{12\rho R}{4\pi R^2 \sigma'} \sum_{i=1}^{3} \int_S \frac{(1 - \delta_{ij})a_{i1}(\hat{s}_j u_i(P) - \hat{s}_i u_j(P)) + \delta_{ij} a_{j1}(\hat{s}, \mathbf{u}(P))}{W^4} \, dS \tag{6.41}$$

where $\sigma' = 1 + 2/\alpha$, a_{ij} are the elements of the matrix A in (6.27), *and* $(\hat{s}, \mathbf{u})$ *is the scalar product.*

Proof. It is sufficient to give the proof for $j = 1$. Using Theorem 6.4, we can rewrite each integral as a series of operators and then, using the properties of the solution, contract the series. Let us consider the integral

$$I = \sum_{i=3}^{3} \int_S \frac{\hat{s}_i \hat{s}_1}{W^5} u_i(P)\, dS.$$

Substituting the representations of the integrals from Theorem 6.4, we have

$$I = I_0 + \sum_{p=4}^{\infty} \frac{(2\rho)^{p-4}}{2R} \frac{(-1)^p}{(2p-1)!!} \sum_{i=1}^{3} \Big[\sum_{k=1}^{3} \sum_{l=1}^{3} a_{1k} a_{il} \frac{\partial}{\partial n_k} \frac{\partial}{\partial n_l} \Big] \frac{\partial^{p-4} \Delta u_i(M)}{\partial n_1^{p-4}}$$

where I_0 is the sum of the first terms with $p \leq 3$. We shall direct our attention to the expression in square brackets, which is simply the notation of the partial derivatives $(\partial/\partial x_1)(\partial/\partial x_i)$ in the coordinates (n_1, n_2, n_3). If we apply this mixed derivative to Δu_i and perform summation over i as written above, we obtain the first component of the vector grad div $\Delta \mathbf{u}$, which is equal to zero for the solution $\mathbf{u}$ by virtue of (6.39). As a result, only I_0 does not vanish:

$$I = a_{11}(A^T \mathbf{u})_1 \frac{5R^2 + 7\rho^2}{15R(R^2 - \rho^2)^3} + a_{21}(A^T \mathbf{u})_2 \frac{5R^2 - \rho^2}{15R(R^2 - \rho^2)^3}$$

$$+ a_{31}(A^T \mathbf{u})_3 \frac{5R^2 - \rho^2}{15R(R^2 - \rho^2)^3} + \frac{(1 - 2/\alpha)\Delta u_1}{30R(R^2 - \rho^2)}$$

$$+ \frac{2\rho}{15R(R^2 - \rho^2)^2} \Big\{ a_{11} \operatorname{div} \mathbf{u} - \frac{\partial u_1}{\partial n_1} + a_{11} \frac{\partial u_1}{\partial x_1} + a_{21} \frac{\partial u_2}{\partial x_1} + a_{31} \frac{\partial u_3}{\partial x_1} \Big\}. \tag{6.42}$$

Using

$$\frac{\partial u_1}{\partial n_i} = a_{1i} \frac{\partial u_1}{\partial x_1} + a_{2i} \frac{\partial u_1}{\partial x_2} + a_{3i} \frac{\partial u_1}{\partial x_3},$$

the expression in the braces in (6.42) can be rewritten as

$$a_{11} \operatorname{div} \mathbf{u} + a_{21} (\operatorname{rot} \mathbf{u})_{21} + a_{31} (\operatorname{rot} \mathbf{u})_{31}, \tag{6.43}$$

where

$$(\operatorname{rot} \mathbf{u})_{ij} = \frac{\partial u_i}{\partial x_j} - \frac{\partial u_j}{\partial x_i}.$$

It remains to give the expressions for div**u** and for the vector rot**u** in terms of surface integrals:

$$\frac{1}{4\pi R^2}\sum_{i=1}^{3}\int\limits_S \frac{s_i}{W^4}(A^T\mathbf{u})_i\, dS = \frac{2\rho}{3R(R^2-\rho^2)^2}(A^T\mathbf{u})_1 + \frac{\operatorname{div}\mathbf{u}}{3R(R^2-\rho^2)},$$

$$\frac{1}{4\pi R^2}\sum_{i=1}^{3}\int\limits_S \frac{s_i}{W^4}(a_{1i}u_2 - a_{2i}u_1)\, dS$$
$$= (a_{11}u_2 - a_{21}u_1)\frac{2\rho}{3R(R^2-\rho^2)^2} + \left\{\frac{\partial u_2}{\partial x_1} - \frac{\partial u_1}{\partial x_2}\right\}\frac{1}{3R(R^2-\rho^2)}$$

and

$$\frac{1}{4\pi R^2}\sum_{i=1}^{3}\int\limits_S \frac{s_i}{W^4}(a_{1i}u_3 - a_{3i}u_1)\, dS$$
$$= (a_{11}u_3 - a_{31}u_1)\frac{2\rho}{3R(R^2-\rho^2)^2} + \left\{\frac{\partial u_3}{\partial x_1} - \frac{\partial u_1}{\partial x_3}\right\}\frac{1}{3R(R^2-\rho^2)}.$$

The last three formulae are obtained on the basis of (6.39), (6.40) and by Theorem 6.4. Deducing div**u** and rot**u** from these formulae and substituting them into (6.43) and then into (6.42), we obtain the desired result. □

In addition to **u**, the formula (6.41) in Theorem 6.5 contains also $\Delta\mathbf{u}$. Therefore, let us develop another auxiliary mean value relation on the basis of property (6.38). This relation is of interest not only as a step on the road to the solution of the main problem, but as a relation in its own, too.

Theorem 6.6.

Suppose we have a function $u(x)$ which is biharmonic in a ball and

$$u = \varphi_0(y), \quad \Delta u = \varphi_1(y), \quad y \in S(0,R), \quad y = (R, \theta', \varphi').$$

Then the value of the function u at an arbitrary internal point $x = (\rho, \theta, \varphi)$ is expressed in terms of $\varphi_0(y)$ and $\varphi_1(y)$:

$$u(x) = -\frac{R^2-\rho^2}{16\pi R^2 k^{3/2}} \int\limits_{S(0,R)} \varphi_1(y) \int\limits_0^k \frac{\sqrt{t}(1-t^2)}{W_t^3}\, dt\, dS(y) + \frac{(R^2-\rho^2)R}{4\pi R^2} \int\limits_{S(0,R)} \frac{\varphi_0(y)}{W^3}\, dS(y) , \tag{6.44}$$

where $W_t^2 = 1 + t^2 - 2t\alpha_1$, $k = \rho/R$.

Proof. We take the biharmonic function u in the form:

$$u(x) = (R^2 - \rho^2)v(x) + w(x) \tag{6.45}$$

where v and w are harmonic functions for which we can pose the boundary value problems:

$$\begin{cases} \Delta w(x) = 0, & x \in B(0,R), \\ w\Big|_S = \varphi_0(y), \end{cases} \tag{6.46}$$

and

$$\begin{cases} \Delta v(x) = 0, & x \in B(0,R), \\ \dfrac{\partial v}{\partial r} + \dfrac{3}{2r}v\Big|_S = -\dfrac{\varphi_1(y)}{4R} . \end{cases} \tag{6.47}$$

Indeed, operating by Δ in (6.45) we find

$$\Delta u(x) = -6v(x) - \sum_{i=1}^{3} 4x_i \frac{\partial v}{\partial x_i} + (R^2-\rho^2)\Delta v(x) + \Delta w(x) .$$

Since the functions v and w are harmonic we get

$$\Delta u(x) = -6v(x) - 4\rho\frac{\partial v(x)}{\partial \rho} .$$

Hence on the boundary we have

$$\varphi_1(y) = -6v(y) - 4\rho\frac{\partial v(y)}{\partial \rho} .$$

We can immediately write down the Poisson integral formula for the function $w(x)$:

$$w(x) = \frac{R(R^2 - \rho^2)}{4\pi R^2} \int_S \frac{\varphi_0(y)}{W^3}\, dS(y)\,. \tag{6.48}$$

Let us derive a similar relation for $v(x)$. For simplicity, let us write $v(\rho, \theta, \varphi)$ for the value $v(x)$ at the point $x = (\rho, \theta, \varphi)$. Using the method of separation of variables, we can write a solution to (6.47) as a series in spherical functions [3] where it is supposed that the function φ_1 is expanded into the Fourier series

$$v(\rho, \theta, \varphi) = -\frac{1}{4}\sum_{n=0}^{\infty} k^n \frac{2}{2n+3} Y_n(\theta, \varphi), \tag{6.49}$$

where

$$Y_n(\theta, \varphi) = \sum_{l=0}^{n} (A_{nl}\cos l\varphi + B_{nl}\sin l\varphi) P_n^{(l)}(\cos\theta), \tag{6.50}$$

$$A_{00} = \frac{1}{4\pi}\int_0^{2\pi}\int_0^{\pi} \varphi_1(\theta', \varphi')\sin\theta'\, d\theta'\, d\varphi',$$

$$A_{nl} = \frac{(2n+1)(n-l)!}{2\pi(n+l)!}\int_0^{2\pi}\int_0^{\pi} \varphi_1(\theta', \varphi') P_n^{(l)}(\cos\theta')\cos l\varphi' \sin\theta'\, d\theta'\, d\varphi',$$

for $n > 0$, and

$$B_{nl} = \frac{(2n+1)(n-l)!}{2\pi(n+l)!}\int_0^{2\pi}\int_0^{\pi} \varphi_1(\theta', \varphi') P_n^{(l)}(\cos\theta')\sin l\varphi' \sin\theta'\, d\theta'\, d\varphi'.$$

The values of the function v are not changed if we introduce a parameter t and integrate (6.49) as follows:

$$v(\rho, \theta, \varphi) = -\frac{1}{4k^{3/2}}\int_0^{k} t^{1/2}\sum_{n=0}^{\infty} t^n Y_n\, dt. \tag{6.51}$$

The functional series in (6.51) can be integrated term by term, since we have an absolutely and uniformly convergent series ($t < 1$ always holds). The resultant series is easy to contract [6], so we get

$$v(\rho,\theta,\varphi) = -\frac{R}{16\pi\rho^{3/2}}\int_0^{\rho}\sqrt{t_1}\int_0^{2\pi}\int_0^{\pi}\varphi_1(R,\theta',\varphi')\frac{R^2-t_1^2}{W_{t_1}^3}\ \sin\theta' d\theta' d\varphi' dt_1. \tag{6.52}$$

Adding properly weighted expressions for v and w in (6.48) and (6.52), we arrive at the desired relation. □

Corollary 6.2.

If $k = 0$, then (6.44) reduces to a known mean value relation for biharmonic functions at the center of the ball:

$$u(0,\theta,\varphi) = \frac{1}{4\pi}\Big\{\int_{\Omega}\varphi_0\ d\Omega - \frac{R^2}{6}\int_{\Omega}\varphi_1\ d\Omega\Big\}.$$

To prove this, it is sufficient to take the limit in (6.44) as $k \to 0$.

Theorem 6.7.

The components of the displacement vector in the Lamé equation satisfy the following mean value relation at an arbitrary interior point in the ball:

$$u_j(x) = \frac{1}{4\pi R^2}\int_S k_{ji}(x,y)u_i(y)\ dS(y)\ , \tag{6.53}$$

$$\begin{aligned} k_{ji}(x,y) &= \frac{\delta_{ij}R(R^2-\rho^2)}{W^3} \\ &+ \frac{R(R^2-\rho^2)\rho^{(1-3\sigma')/(2\sigma')}}{\sigma'}\int_0^{\rho}\frac{\Phi_1(t_1)+\Phi_2(t_1)+\Phi_3(t_1)}{t_1^{(1-\sigma')/(2\sigma')}}\ dt_1\ , \end{aligned}$$

where

$$\Phi_1(t_1) = \frac{\delta_{ij}}{2W_{t_1}^3}\ ,$$

$$\Phi_2(t_1) = -3t_1\frac{(a_{i1}\hat{s}_j + a_{j1}\hat{s}_i) - \delta_{ij}[a_{k1}\hat{s}_k]}{W_{t_1}^4}\ ,$$

and

$$\Phi_3(t_1) = \frac{3(5\hat{s}_i\hat{s}_j - 2\delta_{ij})(R^2 - t_1^2)}{2W_{t_1}^5} .$$

Here we used the notation $[a_{k1}\hat{s}_k] = \sum_{k=1}^{3} a_{k1}\hat{s}_k$.

Proof. The displacement vector $\mathbf{u}$ is termwise biharmonic; therefore we can take advantage of the previous theorem. Let us consider (6.52). The inner integral is the Poisson kernel in $\mathbb{R}^3$ for the harmonic function Δu_i, (since $\Delta(\Delta u_i) = 0$):

$$v_j(\rho, \theta, \varphi) = -\frac{1}{4\rho^{3/2}} \int_0^{\rho} \Delta u_j(t_1, \theta, \varphi)\sqrt{t_1}\, dt_1. \tag{6.54}$$

If we substitute the values of $\Delta u_j(t_1, \theta, \varphi)$ from (6.41) into (6.54) and solve the resultant equation for v_j, we prove the theorem. Thus, for $j = 2$ (other versions are proved in a similar manner) the formula (6.41) is

$$\Delta u_2(t_1, \theta, \varphi) = -\frac{2u_2(t_1, \theta, \varphi)}{(R^2 - t_1^2)\sigma'} + f_1(t_1, \theta, \varphi) ,$$

where the sum of all the integrals is designated by the function f_1. Now we substitute the last mentioned expression in (6.54) and obtain

$$v_2(\rho, \theta, \varphi) = -\frac{1}{4\rho^{3/2}} \int_0^{\rho} \sqrt{t_1} f_1(t_1, \theta, \varphi)\, dt_1 + \frac{1}{2\rho^{3/2}\sigma'} \int_0^{\rho} \frac{\sqrt{t_1}u_2}{R^2 - t_1^2}\, dt_1.$$

By virtue of (6.45) and (6.48), the second term can be transformed, namely

$$\begin{aligned} v_2(\rho, \theta, \varphi) = \; & -\frac{1}{4\rho^{3/2}} \int_0^{\rho} \sqrt{t_1} f_1(t_1, \theta, \varphi)\, dt_1 + \frac{1}{2\rho^{3/2}\sigma'} \int_0^{\rho} \sqrt{t_1} v_2(t_1, \theta, \varphi)\, dt_1 \\ & + \frac{R}{8\rho^{3/2}\sigma'\pi R^2} \int_0^{\rho} \int_S \frac{u_2\sqrt{t_1}}{W_{t_1}^3}\, dt_1\, dS. \end{aligned}$$

Let us introduce the notation

$$f = -f_1 + \frac{1}{2\pi\sigma' R} \int\limits_S \frac{u_2}{W_{t_1}^3}\, dS \ ;$$

then

$$\rho^{3/2} v_2(\rho,\theta,\varphi) - \frac{1}{2\sigma'} \int\limits_0^\rho \sqrt{t_1}\, v_2(t_1,\theta,\varphi)\, dt_1 = \frac{1}{4} \int\limits_0^\rho \sqrt{t_1}\, f(t_1)\, dt_1 .$$

After we differentiate both sides of this equality with respect to ρ and collect the similar terms, we obtain

$$v_2'(\rho,\theta,\varphi) + \frac{1}{\rho}\Big(\frac{3}{2} - \frac{1}{2\sigma'}\Big) v_2(\rho,\theta,\varphi) = \frac{1}{4\rho} f(\rho,\theta,\varphi).$$

This is an ordinary differential equation. Its general solution is

$$v_2(\rho,\theta,\varphi) = \frac{1}{4}\rho^{(1-3\sigma')/(2\sigma')} \int\limits_0^\rho \frac{f(t_1,\theta,\varphi)}{t_1^{(1-\sigma')/(2\sigma')}}\, dt_1 \ .$$

Integrating in new variables $t_1 = tR$ we can rewrite

$$v_2(\rho,\theta,\varphi) = \frac{1}{4} k^{(1-3\sigma')/(2\sigma')} \int\limits_0^k \frac{f(tR,\theta,\varphi)}{t^{(1-\sigma')/(2\sigma')}}\, dt \ .$$

By definition, $\sigma' = 1 + 2/\alpha$ and $\alpha = 1 + \lambda/\mu$; hence, $\sigma' > 0$, $t < 1$, and the integral has no singularities, except when $k = 0$. At $k = 0$ it has a removable singularity to be considered below.

We have thus expressed $v_i(\rho,\theta,\varphi)$ and $w_i(\rho,\theta,\varphi)$ in terms of their values on the sphere, and the theorem is proved, because $u_i = (R^2 - \rho^2)v_i + w_i$. □

To verify (6.45), let us compare it with the known value for $u_i(0,\theta,\varphi)$.

Corollary 6.3.

In the limit, as $k \to 0$, the above mean value relation becomes a known relation for the system of the Lamé equations at the center of the ball:

$$u_i(0) = \frac{3}{8\pi(3+\alpha)} \int_{\Omega} \{(2-\alpha)u_i + 5\alpha s_i s_j u_j\}\, d\Omega.$$

Proof.

When taking the limit $\rho \to 0$ in (6.53), we have to check those integrands whose numerators contain t_1 to the power $(\sigma'-1)/(2\sigma')$. The other functions contain t_1 to higher powers, and therefore their integrals are equal to zero in the limit, as $\rho \to 0$.

Consider the integral

$$\frac{1}{k^{(3\sigma'-1)/(2\sigma')}} \int_0^k \frac{t^{(\sigma'-1)/(2\sigma')}}{(1+t^2-2t\cos\alpha)^{3/2}}\, dt \qquad (6.55)$$

and rewrite its kernel in the form

$$\frac{t^{(\sigma'-1)/(2\sigma')}}{W_t^3} = t^{(\sigma'-1)/(2\sigma')} + t^{(\sigma'-1)/(2\sigma')} \frac{tC_3}{(1+t^2-2t\cos\alpha)^{-3/2}+1},$$

where

$$C_3 = \sum_{i=1}^{3} C_3^i t^{i-1} (t - 2\cos\alpha)^i.$$

Then the integral in (6.55) is

$$k^{(1-3\sigma')/(2\sigma')} \int_0^k t^{(\sigma'-1)/(2\sigma')} + t^{(\sigma'-1)/(2\sigma')} \frac{tC_3}{W_t^{-3}+1}\, dt$$

$$= \frac{2\sigma'}{3\sigma'-1} + k^{(1-3\sigma')/(2\sigma')} \int_0^k \frac{t^{(3\sigma'-1)/(2\sigma')} C_3}{W_t^{-3}+1}\, dt.$$

The other two similar integrals can be transformed in a similar way:

$$k^{(1-3\sigma')/(2\sigma')} \int_0^k \frac{t^{(\sigma'-1)/(2\sigma')}}{W_t^5}\, dt$$

$$= \frac{2\sigma'}{3\sigma'-1} + k^{(1-3\sigma')/(2\sigma')} \int_0^k \frac{t^{(3\sigma'-1)/(2\sigma')} C_5}{W_t^{-5}+1}\, dt,$$

$$k^{(1-3\sigma')/(2\sigma')} \int_0^k \frac{t^{(\sigma'-1)/(2\sigma')}}{W_t^7}\, dt$$

$$= \frac{2\sigma'}{3\sigma'-1} + k^{(1-3\sigma')/(2\sigma')} \int_0^k \frac{t^{(3\sigma'-1)/(2\sigma')} C_7}{W_t^{-7}+1}\, dt \,,$$

where

$$C_q = \sum_{i=1}^{q} C_q^i t^{i-1} (t - 2\cos\alpha)^i, \quad q = 5, 7.$$

Now let us take the limit in (6.53) as $k \to 0$. We obtain

$$u_j(0,\theta,\varphi) = \frac{3}{(3\sigma'-1)\omega_3} \int_\Omega (5\hat{s}_i\hat{s}_j + (\sigma'-2)\delta_{ij}) u_i \, d\Omega.$$

Putting $\sigma' = 1 + 2/\alpha$, we obtain the desired results.

6.3. An alternative derivation of the Poisson formula

In Sections 6.1 and 6.2 we have derived the generalized Poisson formula for the Lamé equation starting from the functional power expansion. In this section we show that the proof given in [21] and [23] leads to the same mean value relation.

Let us consider the first boundary value problem for the Lamé equation in the ball

$$\begin{aligned} \Delta^* \mathbf{u}(x) &= 0, \quad x \in B(0,R)\,, \\ \mathbf{u}(y) &= \mathbf{f}(y), \quad y \in S(0,R)\,. \end{aligned} \tag{6.56}$$

A general representation of the solution to (6.56) is already known in the form of a power series (see, for example, [23]). Under some general assumptions about the function $\mathbf{f}$ (for instance, it should be assumed that $\mathbf{f}$ can be expanded into a Fourier series in spherical functions: $f_i(y) = \sum_{n=0}^{\infty} A_n^i(y)$) this representation has the form

$$u_i(x) = \sum_{n=0}^{\infty} A_n^i \left(\frac{r}{R}\right)^n + (r^2 - R^2) \sum_{n=0}^{\infty} D_{n+2} \frac{\partial d_{n+2}}{\partial x_i} \,, \quad i = 1,2,3, \tag{6.57}$$

where $x = (r, \theta, \varphi)$,

$$D_n = -\frac{\lambda + \mu}{2[\lambda(n-1) + \mu(3n-2)]}, \quad d_n = \sum_{i=1}^{3} \frac{\partial}{\partial x_i} \left\{ A_n^i \left(\frac{r}{R}\right)^n \right\},$$

and A_n^i are the standard spherical functions of order n, i.e.,

$$A_n^i(\theta, \varphi) = \frac{2n+1}{4\pi R^2} \int\limits_{S(0,R)} P_n(\cos\gamma) f_i(\theta, \varphi)\, dS_y ,$$

where $P_n(\xi)$ is the Legendre polynomial of order n and γ is the angle between the vectors x and y:

$$\cos\gamma = \frac{1}{rR} \sum_{k=1}^{3} x_k y_k .$$

The derivation of the expansion given by Thomson is reported in [23]. A different approach to get (6.57) based on potential theory is given in [21]. Using this representation, it is possible to obtain the spherical mean value relation. This was done in [21]. Let us formulate the relevant statement given in [21].

Theorem 6.8.

Under the assumptions $f_i \in C(S(0,R))$, the unique regular solution to the problem (6.56) *satisfies the spherical mean value relation*

$$u_i(x) = \int\limits_{S(0,R)} k_{ij}(x,y) f_j(y)\, dS(y) , \tag{6.58}$$

where

$$k_{ij} = \frac{1}{4\pi R} \left\{ \delta_{ij} \frac{R^2 - \rho^2}{|x-y|^3} + \beta_1 (R^2 - \rho^2) \frac{\partial^2 \Phi(x,y)}{\partial x_i \partial x_j} \right\},$$

$$\Phi(x,y) = \int_0^1 \left\{ \frac{R^2 - \rho^2 t^2}{(R^2 - 2R\rho t\cos\gamma + \rho^2 t^2)^{3/2}} - \frac{1}{R} - \frac{3\rho t \cos\gamma}{R^2} \right\} \frac{dt}{t^{1+\alpha_1}},$$

$$\beta_1 = \frac{\lambda+\mu}{2(\lambda+3\mu)}, \quad \alpha_1 = \frac{\lambda+2\mu}{\lambda+3\mu} \,.$$

Here γ is the angle between the vectors x and y.

Proof. The double layer potential representation reads

$$\mathbf{u}(x) = \int\limits_{S(0,R)} T^*(x-y, n(y))\mathbf{g}(y)\, dS(y), \quad x \in B(0,R) \,, \tag{6.59}$$

where

$$T^*_{kj}(x-y, n(y)) = \frac{1}{4\pi}\left\{\delta_{kj}A\frac{\partial}{\partial n(y)}\frac{1}{r} + 3B\frac{\partial r}{\partial x_k}\frac{\partial r}{\partial x_j}\frac{\partial}{\partial n(y)}\frac{1}{r} - AM_{kj}\Big(\frac{\partial}{\partial y}, n(y)\Big)\frac{1}{r}\right\} \,,$$

$$A = \frac{2\mu}{\lambda+2\mu}, \quad B = \frac{2(\lambda+\mu)}{\lambda+2\mu} \,,$$

and

$$\frac{\partial}{\partial n(x)} \equiv \frac{\partial}{\partial \rho}, \quad \rho = \sqrt{x_1^2+x_2^2+x_3^2} \,, \quad r^2 = \sum_{i=1}^{3}(x_i-y_i)^2 \,,$$

$$M_{kj}\Big(\frac{\partial}{\partial x}, n(x)\Big) = n_j(x)\frac{\partial}{\partial x_k} - n_k(x)\frac{\partial}{\partial x_j} = \frac{1}{\rho}\Big(x_j\frac{\partial}{\partial x_k} - x_k\frac{\partial}{\partial x_j}\Big) \,.$$

The unknown density $\mathbf{g}$ satisfies the boundary integral equation

$$-\mathbf{g}(z) + \int\limits_{S(0,R)} T^*(z-y, n(y))\mathbf{g}(y)\, dS(y) = \mathbf{f}(z) \,. \tag{6.60}$$

The density $\mathbf{g}$ is constructed in the form of a series in standard spherical functions [21] and from (6.59), (6.60) it follows that

$$\mathbf{u}(x) = \sum_{m=0}^{\infty}\Big(\frac{\rho}{R}\Big)^m A_m + (R^2-\rho^2)\sum_{m=2}^{\infty}\frac{\lambda+\mu}{2[m(\lambda+3\mu)-\lambda-2\mu]}\operatorname{grad}\operatorname{div}\left[\Big(\frac{\rho}{R}\Big)^m A_m\right] \,, \tag{6.61}$$

where $A_m = (A_m^1, A_m^2, A_m^3)$.

Standard contracting procedure applied to the series (6.61) yields the desired result. □

An analogous approach is used in [2] to solve the same problem in $2D$.

Remark 6.1.

Note that if we take the second derivatives of the function $\Phi(x, y)$ in the kernel of (6.58) we will come to the spherical mean value relation which we obtained in (6.7). Indeed,

$$\frac{\partial^2 \Phi}{\partial x_i \partial x_j} = \int_0^1 \left\{ 15 \frac{(R^2 - \rho^2 t^2)\hat{s}_i \hat{s}_j}{Q^5} - \frac{2\delta_{ij}}{Q^3} \right.$$
$$\left. - \frac{6(x_i \hat{s}_j + x_j \hat{s}_i)}{Q^4} - \frac{3(R^2 - \rho^2 t^2)\delta_{ij}}{Q^5} \right\} \frac{dt}{t^{\alpha_1 - 1}} ,$$

where $Q = (R^2 - 2R\rho t \cos\gamma + \rho^2 t^2)^{1/2}$. From this we come to (6.7) if we take the relation $t x_k = y_k - Q \hat{s}_k$ into account and use the change of variables $t_1 = \rho t$.

The same technique can be used to derive a spherical mean value relation for the second boundary value problem

$$\begin{aligned} \Delta^* \mathbf{u}(x) &= 0, \quad x \in B(0, R) , \\ T\mathbf{u}(y) &= \mathbf{f}(y), \quad y \in S(0, R) , \end{aligned} \tag{6.62}$$

where the operator T is defined by

$$T_{kj} = \lambda n_k(y) \frac{\partial}{\partial y_j} + \mu n_j(y) \frac{\partial}{\partial y_k} + \mu \delta_{kj} \frac{\partial}{\partial n(y)} .$$

For details, see [21].

Theorem 6.9.

The unique (to within two arbitrary constants C and C_1) regular solution to the problem (6.62) satisfies the following spherical mean value relations

$$\mathbf{u}(x) = \int_{S(0,R)} \mathcal{K}(x,y)\mathbf{f}(y)\, dS(y) + C + |xC_1| ,$$

where C, C_1 *are some arbitrary constants, and*

$$T\mathbf{u}(x) = \frac{1}{4\pi\rho} \int_{S(0,R)} \mathcal{K}^1(x,y)\mathbf{f}(y)\, dS(y) ,$$

where

$$\mathcal{K}_{ij}(x,y) = \frac{1}{8\pi\mu}\Bigg\{(\Phi_1+\Phi_2)\delta_{ij} + \frac{R^2-3\rho^2}{2}\frac{\partial^2\Phi_3}{\partial x_i \partial x_j} + x_j\left(\frac{\partial\Phi_1}{\partial x_i} - \frac{\partial\Phi_2}{\partial x_i}\right) - 2x_i\frac{\partial\Phi_1}{\partial x_j} + x_i\frac{\partial}{\partial x_j}\left(2\rho\frac{\partial\Phi_3}{\partial\rho} - \Phi_3\right) + \rho^2\left(\frac{\partial^2\Phi_2}{\partial x_i\partial x_j} - \frac{\partial^2\Phi_1}{\partial x_i\partial x_j}\right)\Bigg\}$$

and

$$\mathcal{K}^1_{ij}(x,y) = \frac{R^2-\rho^2}{|x-y|^3}\delta_{ij} + (R^2-\rho^2)\frac{\partial^2\Phi_4(x,y)}{\partial x_i\partial x_j} .$$

Here

$$\Phi_1 = \int_0^1 \left[\frac{R^2-\rho^2t^2}{Q(t)} - \frac{1}{R}\right]\frac{dt}{t} , \quad \Phi_2 = \int_0^1 \left[\frac{R^2-\rho^2t^2}{Q(t)} - \frac{1}{R} - \frac{3\rho t\cos\gamma}{R^2}\right]\frac{dt}{t^2} ,$$

$$\Phi_3 = b_1^{-1}\mathrm{Im}\,\Phi_0, \quad \Phi_0 = \int_0^1 \left[\frac{R^2-\rho^2t^2}{Q(t)} - \frac{1}{R}\right]\frac{dt}{t^{1+\alpha_1}} , \quad \Phi_4 = \mathrm{Re}\,(b_2\Phi_0) ,$$

$$\alpha_1 = b_0 + ib_1 = \frac{\mu + i\sqrt{2\lambda^2+6\lambda\mu+3\mu^2}}{2(\lambda+\mu)} , \quad b_2 = \frac{1}{2} + i\frac{3\lambda+4\mu}{2\sqrt{2\lambda^2+6\lambda\mu+3\mu^2}},$$

and

$$Q(t) = (R^2 - 2R\rho t\cos\gamma + \rho^2t^2)^{3/2} .$$

Chapter 7

Spherical means for the stress and strain tensors

In this chapter we derive spherical mean value relations using the generalized Gauss formula (see [53]). This approach was used in [6] to deduce the mean value relation for the Lamé equation. Note that this approach directly gives the mean value relation for an arbitrary sphere $S(x, r) \subset \bar{G}$ under some smoothness assumptions. In Section 7.1 we use this method to obtain the spherical mean value relation deduced in Section 5.2 by the series expansion. In Section 7.2 a mean value theorem which relates the displacements with the stress and strain components is presented. In the next section (Section 7.3) the spherical means of the stress components are related to the values of these components at the center of the sphere.

Note that throughout this chapter we deal with the spherical mean value theorems which relate the spherical means of solutions to the solutions at the center of a ball $B(0, R)$. Therefore, we will simply write, say, u instead of $u(0)$. It will be convenient also to use in this chapter the notation $\operatorname{sh} x$ and $\operatorname{ch} x$ for the functions $\sinh(x)$ and $\cosh(x)$, respectively. and the summation convention (summation over the repeated index), for instance, $u_{i,i}$ means $\operatorname{div} \mathbf{u}$.

7.1. Spherical means for the displacement components through the displacement vector

As in Section 5.2, we consider the equation

$$\mu \Delta \mathbf{u} + (\lambda + \mu) \operatorname{grad} \operatorname{div} \mathbf{u} - \nu \mathbf{u} = 0, \tag{7.1}$$

where $\mathbf{u} = (u_1, u_2, u_3)$ is a real-valued regular vector function. The coefficients λ, μ and ν are assumed to be constants. In this chapter we suppose that ν satisfies the following assumption:

$$(A) \qquad k_s r < \pi, \quad k_p r < \pi \ ,$$

where $k_s = \sqrt{\nu/\mu}$, $k_p = \sqrt{\nu/(\lambda+2\mu)}$.

We multiply (7.1) by r^2 and integrate the result over the ball $B(R) = B(0,R)$, $0 < r < R$:

$$k_s^2 \int\limits_{B(R)} r^2 u_i \, dV = \int\limits_{B(R)} r^2 \left[u_{i,jj} + \alpha u_{j,ji}\right] dV.$$

Integrating the right-hand side of this equation by parts we get

$$\begin{aligned} k_s^2 \int\limits_{B(R)} r^2 u_i \, dV &= \int\limits_{S(R)} R^2 \left[u_{i,j}\frac{x_j}{R} + \alpha u_{j,j}\frac{x_i}{R}\right] dS \\ &\quad - 2 \int\limits_{B(R)} \left[x_j u_{i,j} + \alpha x_i u_{j,j}\right] dV. \end{aligned} \tag{7.2}$$

The surface integral in (7.2) can be transformed as follows

$$\begin{aligned} \int\limits_{S} R^2 \left[u_{i,j}\frac{x_j}{R} + \alpha u_{j,j}\frac{x_i}{R}\right] dS &= R^2 \int\limits_{B(R)} \left[u_{i,jj} + \alpha u_{j,ji}\right] dV \\ &= R^2 k_s^2 \int\limits_{B(R)} u_i \, dV. \end{aligned} \tag{7.3}$$

We integrate by parts the volume integral in the right-hand side of (7.2) and use the relation (7.3). This yields

$$\begin{aligned} & k_s^2 \int\limits_{B(R)} r^2 u_i \, dV - R^2 k_s^2 \int\limits_{B(R)} u_i \, dV \\ &= -2 \int\limits_{S(R)} \left[R^2 u_i + \alpha \frac{x_i x_j}{R} u_j\right] dS + 2 \int\limits_{B(R)} (3+\alpha) u_i \, dV. \end{aligned} \tag{7.4}$$

We recall now (see Section 5.2) that $\mathbf{u}$, the solution to (7.1), can be represented as

$$\mathbf{u} = \mathbf{u}^p + \mathbf{u}^s, \tag{7.5}$$

where $\mathbf{u}^p$, $\mathbf{u}^s$ are the functions solving the following equations:

$$(\Delta - k_p^2)\mathbf{u}^p = 0, \quad \operatorname{rot} \mathbf{u}^p = 0, \tag{7.6}$$

$$(\Delta - k_s^2)\mathbf{u}^s = 0, \quad \operatorname{div} \mathbf{u}^s = 0, \tag{7.7}$$

where

$$k_p = \sqrt{\nu/(\lambda + 2\mu)}, \; k_s = \sqrt{\nu/\mu}, \quad u_i \in C^\infty(G). \tag{7.8}$$

We use now the volume mean value relation for the diffusion equations of the type $(\Delta - \kappa)\mathbf{u} = 0$ which were obtained under the assumption (A) in Section 2.2. This yields (as mentioned above, we use short notations u_i^p and u_i^s instead of $u_i^p(0)$ and $u_i^s(0)$, respectively):

$$4\pi R^3 (f(k_p)u_i^p + f(k_s)u_i^s) = \int\limits_{B(R)} u_i \, dV, \tag{7.9}$$

$$4\pi (f_1(k_p)u_i^p + f_1(k_s)u_i^s) = \int\limits_{B(R)} r^2 u_i \, dV, \tag{7.10}$$

where

$$f(k) = \frac{1}{R^3} \int\limits_0^R r^2 \frac{\operatorname{sh} kr}{kr} \, dr = \frac{\operatorname{ch} kR}{(kR)^2} - \frac{\operatorname{sh} kR}{(kR)^3}, \tag{7.11}$$

$$f_1(k) = \int\limits_0^R r^4 \frac{\operatorname{sh} kr}{kr} dr = f(k)(R^5 + 6\frac{R^3}{k^2}) - 2R^5 \frac{\operatorname{sh} kR}{(kR)^3}. \tag{7.12}$$

We note that (7.9), (7.10) are obtained as the sum of the mean value relations for $\mathbf{u}^p$ and $\mathbf{u}^s$, i.e., from

$$4\pi \frac{\operatorname{sh} k_s R}{k_s R} \mathbf{u}^s + 4\pi \frac{\operatorname{sh} k_p R}{k_p R} \mathbf{u}^p = \int\limits_\Omega \mathbf{u} \, d\Omega, \tag{7.13}$$

when multiplying (7.13) by r^2 (by r^4 in the case of (7.10)), and then integrating over r.

The formulae (7.9), (7.10) allow us to escape the volume integrals in (7.4). Indeed,

$$\begin{aligned}&\left[k_s^2 f_1(k_s) - k_s^2 R^2 f_1(k_p) - 2(3+\alpha)R^3 f(k_p)\right] u_i^p \\ &+\left[k_s^2 f_1(k_p) - k_s^2 R^2 f_1(k_s) - 2(3+\alpha)R^3 f(k_s)\right] u_i^s \\ &= -\frac{2}{4\pi}\int\limits_{S(R)} R u_i + \alpha \frac{x_i x_j}{R} u_j \, dS.\end{aligned}$$

Substituting in this equality $f_1(k_p)$, $f_1(k_s)$ and $f(k_s)$, $f(k_p)$ we get the relation obtained in Section 5.2:

$$N^{(1)}\mathbf{u} = W_1 \mathbf{u}^s + W_2 \mathbf{u}^p, \tag{7.14}$$

where W_1, W_2 and $N^1\mathbf{u}$ are

$$\begin{aligned} W_1 &= (1-\beta)\frac{\operatorname{sh} k_s R}{k_s R} + 3\beta f(k_s)\ , \\ W_2 &= (1+2\beta)\frac{\operatorname{sh} k_p R}{k_p R} - 6\beta f(k_p), \quad \beta = \frac{5(\lambda+\mu)}{2(\lambda+4\mu)}\ , \\ N^1\mathbf{u} &= \frac{1}{4\pi}\int\limits_{S(R)} \left((1-\beta)\delta_{ij} + 3\beta \frac{x_i}{R}\frac{x_j}{R}\right) u_j(y)\, dS(y)\ . \end{aligned} \tag{7.15}$$

Now all the necessary results are obtained and we can formulate the main statement.

Theorem 7.1.

Assume that the parameter ν is chosen so that the assumption (A) is true for all $S(x,r) \subset G$. Then the solution to (7.1) satisfies the following spherical mean value relation

$$\mathbf{u}(x) = \frac{(W_1 - W_2)N_r\mathbf{u} + \left(\frac{\operatorname{sh} k_s R}{k_s R} - \frac{\operatorname{sh} k_p R}{k_p R}\right) N^1\mathbf{u}}{\frac{\operatorname{sh} k_s R}{k_s R} W_2 - \frac{\operatorname{sh} k_p R}{k_p R} W_1}\ . \tag{7.16}$$

for any ball $B(x,R) \subset G$.

Proof. Resolving the linear system of equations (7.14), (7.13) and (7.5) we get the desired relation. The theorem is proved. □

7.2. Mean value relation for the stress and strain tensors through the displacement vector

7.2.1. Mean value relation for the strain components

The main results of this section are Theorems 7.2. and 7.3. To prove them we first obtain a set of preliminary statements.

Let us introduce (see, for example, [21]), the strain tensor

$$\varepsilon_{ik} = \frac{1}{2}(u_{i,k} + u_{k,i}). \tag{7.17}$$

The equalities (7.17), (7.5) and (7.14) give

$$\varepsilon_{ik} = \varepsilon^p_{ik} + \varepsilon^s_{ik}, \tag{7.18}$$

$$W_1 u^s_{i,k} + W_2 u^p_{i,k} = \frac{1}{4\pi R^2} \int\limits_{S(R)} \left(a\delta_{ij} + b\frac{x_i}{R}\frac{x_j}{R}\right) u_{j,k}\, dS. \tag{7.19}$$

We multiply (7.19) by R^4 and integrate over R from 0 to ρ:

$$\begin{aligned} &\int\limits_0^\rho W_1 R^4 u^s_{i,k} dR + \int\limits_0^\rho W_2 R^4 u^p_{i,k} dR \\ &= \frac{1}{4\pi} \int\limits_{B(\rho)} (a\delta_{ij} R^2 + b x_i x_j) u_{j,k}\, dV. \end{aligned} \tag{7.20}$$

The same procedure of integration by parts carried out in the previous section gives

$$\begin{aligned} f_2(k_s) u^s_{i,k} + f_2(k_p) u^p_{i,k} &= \frac{\rho^2}{4\pi} \int\limits_{S(\rho)} \left(a\delta_{ij} + b\frac{x_i x_j}{\rho^2}\right) u_j \frac{x_k}{\rho}\, dS \\ &- \frac{1}{4\pi} \int\limits_{B(\rho)} (2a x_k u_i + b x_i u_k + b u_i x_j \delta_{ik})\, dV, \end{aligned} \tag{7.21}$$

where

$$
\begin{aligned}
f_2(k_s) &= \int_0^\rho W_1 R^4 dR \\
&= \left((6a-3b)\frac{\rho^3}{k_s^2} + \rho^5 a\right) f(k_s) - \rho^2(2a-b)\frac{\operatorname{sh} k_s\rho}{k_s^3}\,, \\
f_2(k_p) &= \int_0^\rho W_2 R^4 dR \\
&= \left(6(a+2b)\frac{\rho^3}{k_p^2} + \rho^5(a+b)\right) f(k_p) - 2(a+b)\rho^2\frac{\operatorname{sh} k_s\rho}{k_p^3}\,.
\end{aligned}
\tag{7.22}
$$

The same is true for $u_{k,i}$; therefore,

$$
\begin{aligned}
&\varepsilon^s_{ik} f_2(k_s) + \varepsilon^p_{ik} f_2(k_p) \\
&\quad = \frac{\rho a}{8\pi}\int_{S(\rho)} (u_i x_k + u_k x_i)\, dS + \frac{b}{4\pi\rho}\int_{S(\rho)} x_i x_j x_k u_j\, dS \\
&\quad - \frac{(2a+b)}{8\pi}\int_{B(\rho)} (x_k u_i + x_i u_k)\, dV - \frac{b}{4\pi}\int_{B(\rho)} u_j x_j \delta_{ik}\, dV\,.
\end{aligned}
\tag{7.23}
$$

To derive the desired mean value relation it remains to pass from the volume integrals in these equalities to surface integrals and use the relations (7.18) and (7.19).

Thus let us evaluate the volume integrals in (7.23).

Lemma 7.1.

The following representations are true

$$
\begin{aligned}
&\int_{B(\rho)} u_i x_k\, dV = 4\pi\big(f_3(k_p) u^p_{i,k} + f_3(k_s) u^s_{i,k}\big), \\
&f_3(k) = \int_0^\rho f(k) r^4\, dr = \frac{\rho^3}{k^2}\left(\frac{\operatorname{sh} k\rho}{k\rho} - 3f(k)\right)\,,
\end{aligned}
\tag{7.24}
$$

or

$$f_3(k) = \frac{1}{2}\Big(\rho^5 f(k) - f_1(k)\Big)\ .$$

Proof. The functions u_i^p solve the diffusion equations; hence the derivatives of $u_{i,k}^p$ also solve the equation

$$\Delta u_{i,k}^p - k_p^2 u_{i,k}^p = 0\ .$$

Therefore, $u_{i,k}^p$ satisfy the relations

$$f(k_p)u_{i,k}^p = \frac{1}{4\pi r^3}\int\limits_{B(r)} u_{i,k}^p\, dV.$$

Integrating by parts, the right-hand side yields

$$f(k_p)u_{i,k}^p = \frac{1}{4\pi r^2}\int\limits_{\Omega} u_i^p x_k\, d\Omega. \tag{7.25}$$

Multiplying this relation by r^4 and integrating over r, $0 \le r \le \rho$ gives:

$$u_{i,k}^p \int\limits_0^\rho f(k_p)r^4 dr = \frac{1}{4\pi}\int_{B(\rho)} u_i^p x_k\, dV,$$

or

$$f_3(k_p)u_{i,k}^p = \frac{1}{4\pi}\int_{B(\rho)} u_i^p x_k\, dV. \tag{7.26}$$

The same arguments for $u_{i,k}^s$ lead to

$$f_3(k_s)u_{i,k}^s(0) = \frac{1}{4\pi}\int\limits_{B(\rho)} u_i^s x_k\, dV. \tag{7.27}$$

We conclude by summing the two last equalities. □

Notice that from (7.24), it follows that

$$\int\limits_{B(\rho)} (u_i x_k + u_k x_i)\, dV = 8\pi\big(f_3(k_p)\varepsilon_{ik}^p + f_3(k_s)\varepsilon_{ik}^s\big). \tag{7.28}$$

Remark 7.1.

We will need a volume integral which can be obtained by proceeding as in Lemma 7.1 where the equality (7.25) is multiplied by r^6.

Omitting these simple calculations we write down the result

$$\int\limits_{B(\rho)} r^2 u_i x_k \, dV = 4\pi\big(f_4(k_p)u^p_{i,k} + f_4(k_s)u^s_{i,k}\big),$$

where

$$f_4(k) = \int\limits_0^\rho f(k) r^6 dr$$
$$= -30\frac{\rho^3}{k^4}f(k) + \frac{\rho^5}{k^2}\Big(\frac{\operatorname{sh} k\rho}{k\rho} - 5\frac{\operatorname{ch} k\rho}{(k\rho)^2} + 15\frac{\operatorname{sh} k\rho}{(k\rho)^3}\Big)\,, \tag{7.29}$$

or

$$f_4(k) = \Big(\frac{10}{k^2} + \rho^2\Big) f_3(k) - 2\frac{\rho^5}{k^2} f(k)\,.$$

Then

$$\int\limits_{B(\rho)} r^2 (u_i x_k + u_k x_i)\, dV = 8\pi\big(f_4(k_p)\varepsilon^p_{ik} + f_4(k_s)\varepsilon^s_{ik}\big)\,. \tag{7.30}$$

Lemma 7.2.

The solution to (7.1) *satisfies the following integral relation*

$$\int\limits_{B(\rho)} u_i x_i \, dV = \frac{f_3(k_p)}{f(k_p)\rho^2} \int\limits_{\Omega} u_i x_i \, d\Omega. \tag{7.31}$$

Proof. Introduce the notations

$$v^p = \sum_{i=1}^3 x_i u^p_i, \quad v^s = \sum_{i=1}^3 x_i u^s_i.$$

Then,

$$\int_B u_i x_i \, dV = \int_B (u_i^p x_i + u_i^s x_i) \, dV = \int_B (v^p + v^s) \, dV.$$

Differentiating v^p yields

$$\begin{aligned} \Delta v^p &= 2\mathrm{div}\, \mathbf{u}^p + \sum_{i=1}^{3} x_i \Delta u_i^p \\ &= 2\mathrm{div}\, \mathbf{u}^p + \sum_{i=1}^{3} x_i k_p^2 u_i^p = 2\mathrm{div}\, \mathbf{u}^p + k_p^2 v^p, \end{aligned} \tag{7.32}$$

$$\begin{aligned} \Delta^2 v^p &= 4\mathrm{div}\, \Delta \mathbf{u}^p + \sum_{i=1}^{3} x_i \Delta^2 u_i^p \\ &= 4\mathrm{div}\, k_p^2 \mathbf{u}^p + \sum_{i=1}^{3} x_i k_p^4 u_i^p = 4\mathrm{div}\, \mathbf{u}^p k_p^2 + k^4 v^p. \end{aligned} \tag{7.33}$$

We resolve the two last equations with respect to v^p. This leads to the relation

$$(\Delta - k_p^2)^2 v^p = 0.$$

Then the function $(\Delta - k_p^2) v^p$ satisfies the relation

$$f(k_p)(\Delta v^p - k_p^2 v^p) = \frac{1}{4\pi r^3} \int_{B(r)} (\Delta - k_p^2) v^p \, dV. \tag{7.34}$$

Substituting (7.32) into (7.34) we come to the relation (taking into account that $v^p(0) = 0$)

$$f(k_p) \Delta v^p = \frac{2}{4\pi r^2} \int_{B(r)} \mathrm{div}\, \mathbf{u}^p \, dV \; .$$

Applying the Gauss theorem we get

$$f(k_p) \Delta v^p = \frac{2}{4\pi r^2} \int_{\Omega} u_i^p x_i \, d\Omega = \frac{2}{4\pi r^2} \int_{\Omega} v^p \, d\Omega. \tag{7.35}$$

Multiplying this equality by r^4 and integrating the result over r, $0 \le r \le \rho$ yields

$$f_3(k_p)\Delta v^p = \frac{2}{4\pi}\int\limits_{B(\rho)} v^p\,dV.$$

From this, in view of (7.35), we get

$$f(k_p)\int\limits_{B(\rho)} v^p\,dV = \frac{f_3(k_p)}{\rho^2}\int\limits_{\Omega} v^p\,d\Omega. \tag{7.36}$$

Since $\operatorname{div}\mathbf{u}^s = 0$, the relation of the type (7.32) for v^s gives that

$$\Delta v^s = k_s^2 v^s,$$

and hence

$$\frac{1}{4\pi R^2}\int\limits_{B(\rho)} v_s\,dV = f(k_s)v(0) = 0.$$

The contribution of v^s to the integral (7.31) is equal to zero and hence (7.36) gives the desired result. The lemma is proved. □

Coming back to (7.23) we substitute the volume integrals obtained in the two last Lemmas. This yields

$$\begin{aligned} 2\varepsilon^s_{ik}\big(f_2(k_s) + (2a+b)f_3(k_s)\big) + 2\varepsilon^p_{ik}\big(f_2(k_p) + (2a+b)f_3(k_p)\big) \\ = \frac{\rho^3 a}{4\pi}\int\limits_{\Omega}(u_i x_k + u_k x_i)\,d\Omega + \frac{\rho b}{2\pi}\int\limits_{\Omega} x_i x_j x_k u_j\,d\Omega \\ - \frac{b\delta_{ik}f_3(k_p)}{2\pi\rho^2 f(k_p)}\int\limits_{\Omega} u_j x_j\,d\Omega. \end{aligned} \tag{7.37}$$

From (7.25),

$$\frac{1}{4\pi\rho^2}\int\limits_{\Omega}(u_i x_k + u_k x_i)\,d\Omega = 2f(k_p)\varepsilon^p_{ik} + 2f(k_s)\varepsilon^s_{ik}. \tag{7.38}$$

Thus we have a system of linear equations (7.18), (7.37) and (7.38). Solving this system yields

Theorem 7.2. *Assume that the parameter ν is chosen so that the assumption (A) is true for all $S(x,r) \subset G$. The strain components are*

related to the displacement vector through the following mean value relation

$$\varepsilon_{ik}(0) = \frac{f(k_s) - f(k_p)}{4\pi R^2 A}\Big[R^3 \int_\Omega x_i x_j x_k u_j \, d\Omega - \delta_{ik}\frac{f_3(k_p)}{f(k_p)} \int_\Omega u_j x_j \, d\Omega\Big]$$

$$+ \frac{1}{8\pi R^2 A}\Big[R^5 f(k_p) - 2f_3(k_s) - 3f_3(k_p)\Big] \int_\Omega (u_i x_k + u_k x_i) \, d\Omega, \tag{7.39}$$

where

$$A = R^5 f(k_p) f(k_s) - 3f_3(k_p) f(k_s) - 2f_3(k_s) f(k_p).$$

7.2.2. Mean value relation for the stress components

In the next statement we present the mean value relation for the stress components through the displacement vector.

Theorem 7.3. *Assume that the parameter ν is chosen so that the assumption (A) is true for all $S(x,r) \subset G$. The stress components τ_{ij} at the center of the sphere are related to the integrals of the displacement vector:*

$$\tau_{ik} = \frac{2\mu(f(k_s) - f(k_p))}{4\pi R^2 A}\Big[R^3 \int_\Omega x_i x_j x_k u_j \, d\Omega - \delta_{ik}\frac{f_3(k_p)}{f(k_p)} \int_\Omega u_j x_j \, d\Omega\Big]$$

$$+ \frac{\mu}{4\pi R^2 A}\Big[R^5 f(k_p) - 2f_3(k_s) - 3f_3(k_p)\Big] \int_\Omega (u_i x_k + u_k x_i) \, d\Omega$$

$$+ \frac{\lambda \delta_{ik}}{4\pi R^2} \frac{1}{f(k_p)} \int_\Omega u_j x_j \, d\Omega.$$

where the quantities A and $f_i(k)$ are defined as in Theorem 7.2.

Proof. By the Hooke law,

$$\tau_{ik} = \mu(u_{i,k} + u_{k,i}) + \lambda u_{j,j}\delta_{ik} = 2\mu\varepsilon_{ik} + \lambda\delta_{ik}\,\mathrm{div}\,\mathbf{u}.$$

This shows that it is required to get a mean value relation for the divergence term $u_{i,i}$ and substitute it in (7.39). From (7.6) it follows that

$$(\Delta - k_p^2)\,\mathrm{div}\,\mathbf{u} = 0;$$

hence

$$f(k_p)\,\mathrm{div}\,\mathbf{u}(0) = \frac{1}{4\pi R^3}\int\limits_{B(R)} \mathrm{div}\,\mathbf{u}\,dV.$$

By the Gauss theorem

$$f(k_p)\,\mathrm{div}\,\mathbf{u}(0) = \frac{1}{4\pi R^2}\int\limits_{\Omega} u_i x_k\,d\Omega;$$

thus

$$u_{i,i} = \frac{1}{4\pi R^2}\frac{1}{f(k_p)}\int\limits_{\Omega} u_i x_i\,d\Omega\ . \tag{7.40}$$

Using this representation and the last theorem we get the result. □

Corollary 7.1.

For the Lamé equation ($\nu = 0$) the mean value relation of this theorem becomes

$$\tau_{ik}(0) = \frac{3\mu}{4\pi(5+2\alpha)R}\Bigg\{5(1-\alpha)\int\limits_{\Omega}\frac{u_i x_k + u_k x_i}{R}d\Omega + 35\alpha\int\limits_{\Omega}\frac{x_i x_j x_k}{R^3}u_j\,d\Omega$$
$$+\ (2\alpha^2 - 4\alpha - 5)\delta_{ik}\int\limits_{\Omega} u_j\frac{x_j}{R}\,d\Omega\Bigg\}\ .$$

It is proved by taking the limit as $\nu \to 0$ in the mean value relation of the above theorem.

7.3. Mean value relations for the stress components in terms of the surface tractions

In this section we deduce a spherical mean value relation between the stress components and the surface tractions $\tau_{ik}\,x_k$.

As in the previous section, the main result, Theorem 7.5, will follow from a set of preliminary results.

We need some properties of τ_{ij}.

Theorem 7.4.

For the stress components, the following representation holds

$$\tau_{ij} = \tau_{ij}^{p} + \tau_{ij}^{s}\ , \tag{7.41}$$

where

$$\Delta\tau_{ij}^{p} - k_p^2\tau_{ij}^{p} = 0, \quad \tau_{ij}^{p} = \frac{(\Delta - k_s^2)\tau_{ij}}{k_p^2 - k_s^2}\ ,$$

$$\Delta\tau_{ij}^{s} - k_s^2\tau_{ij}^{s} = 0, \quad \tau_{ij}^{s} = \frac{(\Delta - k_p^2)\tau_{ij}}{k_s^2 - k_p^2}\ ,$$

$$\tau_{ii}^{p} = \sum_{i=1}^{3}\tau_{ii}^{p} = \frac{(\Delta - k_s^2)\tau_{ii}}{k_p^2 - k_s^2} = \tau_{ii}, \quad \tau_{ii}^{s} = 0\ .$$

Proof. It is known [21] that τ_{ij} satisfies the equilibrium equation

$$\tau_{ij,j} - \nu u_i = 0, \tag{7.42}$$

the equation

$$\varepsilon_{ij} = \frac{1}{2\mu}\tau_{ij} - \frac{\lambda\delta_{ij}}{2\mu(3\lambda + 2\mu)}\tau_{mm} \tag{7.43}$$

and the Hooke equation

$$\tau_{ij} = 2\mu\varepsilon_{ij} + \lambda u_{k,k}\delta_{ij}. \tag{7.44}$$

Differentiation of (7.44) yields

$$\varepsilon_{ij,kk} + \varepsilon_{kk,ij} = \varepsilon_{ik,jl} + \varepsilon_{jl,ik}.$$

In view of (7.43) we get, choosing $l = k$,

$$\tau_{ij,kk} + \tau_{kk,ij} - \frac{\lambda}{3\lambda + 2\mu}(\tau_{mm,kk}\delta_{ij} + \tau_{mm,ij}\delta_{kk})$$
$$= \tau_{ik,jk} + \tau_{jk,ik} - \frac{\lambda}{3\lambda + 2\mu}(\tau_{mm,jk}\delta_{ik} + \tau_{mm,ik}\delta_{jk}).$$

Summation over k yields

$$\sum_k \tau_{ij,kk} + \sum_k \tau_{kk,ij} - \frac{\lambda\delta_{ij}}{3\lambda+2\mu}\sum_k \tau_{kk,ll} - \frac{\lambda}{3\lambda+2\mu}\sum_k \tau_{kk,ij} = \sum_k (\tau_{ik,jk} + \tau_{jk,ik}).$$

From this, in view of (7.42), we get that

$$\Delta\tau_{ij} + \frac{2(\lambda+\mu)}{3\lambda+2\mu}\tau_{mm,ij} - \frac{\lambda\delta_{ij}}{3\lambda+2\mu}\Delta\tau_{mm} = \nu\Big(\frac{\partial u_i}{\partial x_j} + \frac{\partial u_j}{\partial x_i}\Big).$$

The right-hand side can be transformed according to (7.43); hence

$$\Delta\tau_{ij} + \frac{2(\lambda+\mu)}{3\lambda+2\mu}\tau_{mm,ij} - \frac{\nu}{\mu}\tau_{ij} - \frac{\lambda\delta_{ij}}{3\lambda+2\mu}\left(\Delta - \frac{\mu}{\nu}\right)\tau_{mm} = 0. \qquad (7.45)$$

We put $i = j$ and sum over i:

$$\sum_i \tau_{mm,ii} = \frac{\nu}{\lambda+2\mu}\tau_{mm}, \qquad (7.46)$$

i.e.,

$$\Delta\tau_{mm} = k_p^2\tau_{mm}.$$

Using (7.46) and (7.45) we get

$$\Delta\tau_{ij} + \frac{2(\lambda+\mu)}{3\lambda+2\mu}\tau_{mm,ij} - \frac{\nu}{\mu}\left(\tau_{ij} + \frac{\lambda\delta_{ij}(\lambda+\mu)}{(3\lambda+2\mu)(\lambda+2\mu)}\tau_{mm}\right) = 0. \quad (7.47)$$

It remains to apply the operator Δ to the expression (7.47) and use the properties (7.46), (7.47):

$$\Delta^2\tau_{ij} + \frac{\nu}{\lambda+2\mu}\cdot\frac{\nu}{\mu}\tau_{ij} - \frac{\nu}{\lambda+2\mu}\Delta\tau_{ij} - \frac{\nu}{\mu}\Delta\tau_{ij} = 0,$$

or

$$\left(\Delta - \frac{\nu}{\lambda+2\mu}\right)\left(\Delta - \frac{\nu}{\mu}\right)\tau_{ij} = 0$$

which proves the theorem. □

It is convenient to introduce the notation $\gamma = \frac{2(\lambda+\mu)}{3\lambda+2\mu}$. Then the compatibility equation (7.47) reads

$$\Delta\tau_{ij} + \gamma\tau_{mm,ij} - k_s^2\tau_{ij} + \delta_{ij}\frac{k_s^2 - 2k_p^2}{2}\gamma\tau_{mm} = 0. \tag{7.48}$$

Multiplying this equality by r^2 , $0 \le r \le \rho$, and integrating over the ball $B(\rho)$ yields

$$\begin{aligned} \int\limits_{B(\rho)} r^2 k_s^2 \tau_{ij}\, dV - \frac{\delta_{ij}\gamma}{2}(k_s^2 - 2k_p^2) \int\limits_{B(\rho)} r^2\tau_{mm}\, dV \\ = \int\limits_{B(\rho)} r^2(\Delta\tau_{ij} + \gamma\tau_{mm,ij})\, dV. \end{aligned} \tag{7.49}$$

Integration by parts of the last integral leads to

$$\begin{aligned} \int\limits_{B(\rho)} r^2(\Delta\tau_{ij} + \gamma\tau_{mm,ij})\, dV &= \rho^2 \int\limits_{S(\rho)} \left(\tau_{ij,k}\frac{x_k}{\rho} + \gamma\tau_{mm,i}\frac{x_j}{\rho}\right) dS \\ &- 2\int\limits_{B(\rho)} r\left(\tau_{ij,k}\frac{x_k}{r} + \gamma\tau_{mm,i}\frac{x_j}{r}\right) dV\ . \end{aligned} \tag{7.50}$$

Using the Gauss theorem and the equality (7.48), we transform the surface integral in (7.50) to

$$\begin{aligned} \rho^2 \int\limits_{S(\rho)} \left(\tau_{ij,k}\frac{x_k}{\rho} + \gamma\tau_{mm,i}\frac{x_j}{\rho}\right) dS &= \rho^2 \int\limits_{B(\rho)} (\tau_{ij,kk} + \gamma\tau_{mm,ij})\, dV \\ &= \rho^2 \int\limits_{B(\rho)} \left(k_s^2\tau_{ij} - \delta_{ij}\gamma\frac{k_s^2 - 2k_p^2}{2}\tau_{mm}\right) dV. \end{aligned}$$

Then (7.49) can be rewritten as

$$\begin{aligned} &\int\limits_{B(\rho)} r^2 k_s^2\tau_{ij}\, dV - \delta_{ij}\gamma\left(\frac{k_s^2}{2} - k_p^2\right) \int\limits_{B(\rho)} r^2\tau_{mm}\, dV \\ &- \rho^2 \int\limits_{B(\rho)} k_s^2\tau_{ij}\, dV + \rho^2\delta_{ij}\gamma\left(\frac{k_s^2}{2} - k_p^2\right) \int\limits_{B(\rho)} \tau_{mm}\, dV \\ &= -2\int\limits_{B(\rho)} (\tau_{ij,k}x_k + \gamma\tau_{mm,i}x_j)\, dV. \end{aligned}$$

Integrating the right-hand side by parts yields

$$\begin{aligned}
&\int\limits_{B(\rho)} r^2 k_s^2 \tau_{ij}\, dV - \delta_{ij}\gamma\Big(\frac{k_s^2}{2} - k_p^2\Big) \int\limits_{B(\rho)} r^2 \tau_{mm}\, dV \\
&- \quad (\rho^2 k_s^2 + 6) \int\limits_{B(\rho)} \tau_{ij}\, dV + \delta_{ij}\gamma\Big(\rho^2\Big(\frac{k_s^2}{2} - k_p^2\Big) - 2\Big) \int\limits_{B(\rho)} \tau_{mm}\, dV \qquad (7.51) \\
&= \quad -2\rho^3 \int\limits_{\Omega} \tau_{ij}\, d\Omega - 2\rho^2\gamma \int\limits_{\Omega} \tau_{mm} \frac{x_i x_j}{\rho}\, d\Omega.
\end{aligned}$$

We pass from the volume integrals to the surface integrals using τ_{ij}.

From Theorem 7.4 it follows that τ_{ij}^p and τ_{ij}^s satisfy the equation

$$\Delta \tau_{ij}^t - k_t^2 \tau_{ij}^t = 0, \quad t = p, s.$$

Hence,

$$f(k_t)\tau_{ij}^t = \frac{1}{4\pi r^3} \int\limits_{B(r)} \tau_{ij}^t\, dV \ ,$$

and

$$f_1(k_t)\tau_{ij}^t = \frac{1}{4\pi} \int\limits_{B(r)} \rho^2 \tau_{ij}^t\, dV \ .$$

Then

$$f(k_p)\tau_{ij}^p + f(k_s)\tau_{ij}^s = \frac{1}{4\pi r^3} \int\limits_{B(r)} \tau_{ij}\, dV \ , \qquad (7.52)$$

and

$$f_1(k_p)\tau_{ij}^p + f_1(k_s)\tau_{ij}^s = \frac{1}{4\pi} \int\limits_{B(r)} \rho^2 \tau_{ij}\, dV \ , \qquad (7.53)$$

where $f(k)$ and $f_1(k)$ are defined in (7.11) and (7.12). We put $i = j$ in the last equalities and sum over i. By the property (7.41), this yields

$$f(k_p)\tau_{mm} = \frac{1}{4\pi r^3} \int\limits_{B(r)} \tau_{mm}\, dV \ , \qquad (7.54)$$

and

$$f_1(k_p)\tau_{mm} = \frac{1}{4\pi}\int\limits_{B(r)} \rho^2 \tau_{mm}\, dV \ . \tag{7.55}$$

Note that the representation for the volume integrals through the stress tensor components can also be derived from the equilibrium equations. This is seen from the Lemma which follows.

Lemma 7.3.

The following relations are true. First,

$$\int\limits_{B(\rho)} \tau_{ij}\, dV = \frac{\rho}{2}\int\limits_{\Omega} (\tau_{ik}x_j x_k + \tau_{jk}x_i x_k)\, d\Omega$$
$$+\frac{4\pi\lambda\delta_{ij}k_s^2 f_3(k_p)}{2(3\lambda+2\mu)}\tau_{mm} - 4\pi\frac{k_s^2}{2}\left(f_3(k_p)\tau_{ij}^p + f_3(k_s)\tau_{ij}^s\right) , \tag{7.56}$$

then,

$$\int\limits_{B(\rho)} \tau_{mm}\, dV = \rho\int\limits_{\Omega} \tau_{il}x_i x_l\, d\Omega - \frac{4\pi\nu}{(3\lambda+2\mu)} f_3(k_p)\tau_{mm} \ , \tag{7.57}$$

next,

$$\frac{2}{4\pi}\int\limits_{B(\rho)} r^2\tau_{ij}\, dV = \frac{\rho^3}{4\pi}\int\limits_{\Omega} (\tau_{ik}x_j x_k + \tau_{jk}x_i x_k)\, d\Omega$$
$$+\frac{\rho^2\lambda\delta_{ij}k_s^2 f_3(k_p)}{3\lambda+2\mu}\tau_{mm} - (4+\rho^2k_s^2)\Big(f_3(k_p)\tau_{ij}^p + f_3(k_s)\tau_{ij}^s\Big) , \tag{7.58}$$

and

$$\frac{1}{4\pi}\int\limits_{B(\rho)} r^2\tau_{mm}\, dV = \frac{\rho^3}{4\pi}\int\limits_{\Omega} \tau_{ik}x_i x_k\, d\Omega - \tau_{mm}\Big(2+\rho^2\frac{\nu}{3\lambda+2\mu}\Big) f_3(k_p) \ . \tag{7.59}$$

Proof. We give the proof of (7.56). The same arguments work for (7.58); (7.57) and (7.59) follow from (7.56) and (7.58), respectively, when putting $j = i$ and taking a sum over j.

Now, in the equality

$$\int_{B(\rho)} \tau_{ij}\, dV = \int_{B(\rho)} (\tau_{ik}x_j)_{,k}\, dV - \int_{B(\rho)} \tau_{ik,k}x_j\, dV \ .$$

we take by parts the first integral in the right-hand side, and transform the second integral using the equilibrium equation (7.42). This yields

$$\int_{B(\rho)} \tau_{ij}\, dV = \int_{S(\rho)} \tau_{ik}x_j \frac{x_k}{\rho}\, dS - \nu \int_{B(\rho)} u_i x_j\, dV.$$

Since $\tau_{ij} = \tau_{ji}$,

$$2\int_{B(\rho)} \tau_{ij}\, dV = \rho \int_{\Omega} (\tau_{ik}x_j x_k + \tau_{jk}x_k x_i)\, d\Omega - \nu \int_{B(\rho)} (u_i x_j + u_j x_i)\, dV.$$

From (7.28) it follows

$$2\int_{B(\rho)} \tau_{ij}\, dV = \rho \int_{\Omega} (\tau_{ik}x_j x_k + \tau_{jk}x_k x_i)\, d\Omega - 8\pi\nu\big(f_3(k_p)\varepsilon^p_{ij} + f_3(k_s)\varepsilon^s_{ij}\big).$$

From this, using (7.43), we get the desired formula (7.56). □

Now, from (7.52) and (7.56) we obtain the first auxiliary mean value relation for τ_{ij}

$$\frac{1}{4\pi}\int_{\Omega} \left(\tau_{ik}\frac{x_j}{\rho}\frac{x_k}{\rho} + \tau_{jk}\frac{x_i}{\rho}\frac{x_k}{\rho}\right) d\Omega + \frac{\lambda k_s^2 \delta_{ij}}{(3\lambda + 2\mu)\rho^3} f_3(k_p)\tau_{mm}$$
$$= \left(\frac{k_s^2}{\rho^3} f_3(k_p) + 2f(k_p)\right)\tau^p_{ij} + \left(\frac{k_s^2}{\rho^3} f_3(k_s) + 2f(k_s)\right)\tau^s_{ij} \ . \tag{7.60}$$

For the function τ_{mm} we find from (7.54) and (7.57) that

$$\frac{\rho}{4\pi}\int_{\Omega}\tau_{il}x_i x_l\,d\Omega = \tau_{mm}\left(\rho^3 f(k_p) + \frac{\nu}{3\lambda+2\mu}f_3(k_p)\right). \tag{7.61}$$

Using the representations (7.52)–(7.55) we get in (7.51) that

$$\begin{aligned}\frac{\rho^3}{4\pi}\int_{\Omega}\tau_{ij}\,d\Omega + \frac{\rho\gamma}{4\pi}\int_{\Omega}\tau_{mm}x_i x_j\,d\Omega \\ = \tau_{ij}^{p}\left(k_s^2 f_3(k_p) + 3\rho^3 f(k_p)\right) + \tau_{ij}^{s}\left(k_s^2 f_3(k_s) + 3\rho^3 f(k_s)\right) \\ - \tau_{mm}\delta_{ij}\gamma\left(\left(\frac{k_s^2}{2} - k_p^2\right)f_3(k_p) - \rho^3 f(k_p)\right).\end{aligned} \tag{7.62}$$

Multiplying (7.62) by ρ, integrating over ρ, $0 \le \rho \le R$, and in view of (7.58), we get:

$$\begin{aligned}\frac{1}{4\pi}\int_{B(R)}\tau_{mm}x_i x_j\,dV = \frac{\tau_{ij}^{p}}{\gamma}\left(\frac{k_s^2}{k_p^2} - 1\right)\left(f_1(k_p) - 3f_3(k_p)\right) \\ + \tau_{mm}\delta_{ij}\left\{\left(\frac{3k_s^2}{2k_p^2} - 2\right)f_3(k_p) - \left(\frac{k_s^2}{2k_p^2} - 1\right)f_1(k_p)\right\}.\end{aligned} \tag{7.63}$$

It remains to evaluate the last volume integral in (7.63).

Lemma 7.4.

The following representation of the volume integral through the surface integrals is true

$$\begin{aligned}\frac{1}{4\pi}\int_{B(\rho)}\tau_{kk}x_i x_j\,dV = \frac{\rho}{4\pi}\int_{\Omega}\tau_{kl}x_k x_i x_j x_l\,d\Omega \\ - \tau_{ij}^{p}\left(2f_3(k_p) - \frac{k_s^2}{2}f_5(k_p) + \frac{k_s^2}{2}f_4(k_p)\right) - \tau_{ij}^{s}\left(2f_3(k_s) + 2k_s^2 f_5(k_s)\right) \\ + \tau_{mm}\delta_{ij}\left(\frac{k_s^2\lambda}{2(3\lambda+2\mu)}f_4(k_p) - \frac{(\lambda+2\mu)k_s^2}{2(3\lambda+2\mu)}f_5(k_p)\right),\end{aligned} \tag{7.64}$$

where $f_4(k)$ is given above by the formula (7.29), and

$$f_5(k) = \frac{\rho^5}{k^2} f(k) - \frac{5}{k^2} f_3(k) \, .$$

Note that

$$f_5(k) = \frac{1}{k^2}(\rho^5 f(k) - 5 f_3(k)) \, ,$$

and

$$f_4(k) = \rho^2 f_3(k) - 2 f_5(k) \, .$$

Proof. We start from the obvious equality

$$\int\limits_{B(R)} \tau_{kk} x_i x_j \, dV = \int\limits_{B(R)} (\tau_{kl} x_k x_i x_j)_{,l} \, dV - \int\limits_{B(R)} (\tau_{ki} x_k x_j + \tau_{kj} x_k x_i) \, dV - \int\limits_{B(R)} \tau_{kl,l} x_k x_i x_j \, dV \, .$$

From this,

$$\int\limits_{B(R)} \tau_{kk} x_i x_j \, dV = R \int\limits_{\Omega} \tau_{kl} x_k x_i x_j \, d\Omega - \int\limits_{B(R)} (\tau_{ki} x_k x_j + \tau_{kj} x_k x_i) \, dV - \int\limits_{B(R)} \tau_{kl,l} x_k x_i x_j \, dV \, . \tag{7.65}$$

To evaluate the volume integrals we proceed as follows. Let us consider the first integral. Multiplying (7.60) by R and integrating over R, $0 \leq R \leq \rho$ yields

$$\frac{1}{4\pi} \int\limits_{B(\rho)} (\tau_{ik} x_j x_k + \tau_{jk} x_i x_k) \, dV = \left(2 f_3(k_p) + k_s^2 f_5(k_p)\right) \tau_{ij}^p + \left(2 f_3(k_s) + k_s^2 f_5(k_s)\right) \tau_{ij}^s - \frac{\lambda k_s^2 \delta_{ij}}{3\lambda + 2\mu} f_5(k_p) \tau_{mm} \, . \tag{7.66}$$

Now let us evaluate the second volume integral in the right-hand side of (7.65). From (7.37), (7.38) and (7.40) we get

$$\varepsilon_{ik}^p (\rho^5 f(k_p) - 3 f_3(k_p)) + 2\varepsilon_{ik}^s f_3(k_s) = \frac{\rho}{4\pi} \int\limits_{\Omega} u_j x_j x_i x_k \, d\Omega - f_3(k_p) \delta_{ik} \sum_{i=1}^{3} \varepsilon_{ii} \, .$$

Multiplying by ρ and integrating over ρ in the interval $[0, R]$ yields

$$\varepsilon^p_{ik}(f_4(k_p)-3f_5(k_p))+2\varepsilon^s_{ik}f_5(k_s)=\frac{1}{4\pi}\int\limits_{B(R)}u_jx_jx_ix_k\,dV-f_5(k_p)\delta_{ik}\sum_{i=1}^{3}\varepsilon_{ii}\,.$$

Using (7.43), we pass from the strain to the stress tensor

$$\tau^p_{ik}(f_4(k_p)-3f_5(k_p))+2\tau^s_{ik}f_5(k_s)-\left(\frac{\lambda f_4(k_p)}{3\lambda+2\mu}-3f_5(k_p)\right)\delta_{ik}\tau_{mm}$$
$$=\frac{\mu}{2\pi}\int\limits_{B(R)}u_jx_jx_ix_k\,dV\,.$$

From this equality and using the equilibrium equation (7.42) in the form

$$\nu\int\limits_{B(R)}u_jx_jx_ix_k\,dV=\int\limits_{B(R)}\tau_{jl,l}x_jx_ix_k\,dV$$

we obtain the desired integral

$$\frac{1}{4\pi}\int\limits_{B(R)}\tau_{jl,l}x_jx_ix_k\,dV=\frac{k_s^2}{2}(f_4(k_p)-3f_5(k_p))\tau^p_{jk}+k_s^2f_5(k_s)\tau^s_{ij}$$
$$-\delta_{ij}\frac{k_s^2}{2}\left(\frac{\lambda}{3\lambda+2\mu}f_4(k_p)-f_5(k_p)\right)\tau_{mm}\,. \tag{7.67}$$

Finally, from (7.65), (7.66) and (7.67) we get (7.64). □

Now we substitute the representation (7.64) into (7.63) to find the second auxiliary mean value relation

$$\frac{1}{4\pi}\int\limits_{\Omega}\tau_{kl}\frac{x_kx_lx_ix_j}{\rho^4}\,d\Omega$$
$$+\ \tau_{mm}\delta_{ij}\left\{-f(k_p)+\frac{1}{\rho^5}\left(\frac{\lambda k_s^2\rho^2}{2(3\lambda+2\mu)}+4\right)f_3(k_p)\right\}$$

$$=\tau^p_{ij}\left(\frac{1}{\rho^5}\left(12+\frac{k_s^2\rho^2}{2}\right)f_3(k_p)-2f(k_p)\right)+\tau^s_{ij}\left(2f(k_s)-\frac{8}{\rho^5}f_3(k_s)\right). \tag{7.68}$$

Theorem 7.5. *Assume that the parameter ν is chosen so that the assumption (A) is true for all $S(x,r) \subset G$. The stress components at the center of the sphere is related to the surface tractions by*

$$\tau_{ij}(0) = \frac{B_2 - A_2}{4\pi\rho^4(A_1B_2 - A_2B_1)} \int_\Omega \tau_{kl}x_kx_ix_lx_j \, d\Omega$$
$$+ \frac{B_1 - A_1}{4\pi\rho^2(A_1B_2 - A_2B_1)} \int_\Omega [\tau_{ik}x_jx_k - \tau_{jk}x_ix_k] \, d\Omega$$
$$+ \delta_{ij}\frac{C_1(B_2 - A_2) + C_2(B_1 - A_1)}{4\pi\rho^2(A_1B_2 - A_2B_1)C_3} \int_\Omega \tau_{kl}x_kx_l \, d\Omega,$$

where

$$A_1 = -2f(k_p) + \left(\frac{12}{\rho^5} + \frac{k_s^2}{2\rho^3}\right) f_3(k_p) \, ,$$
$$B_1 = 2f(k_s) - \frac{8}{\rho^5} f_3(k_s),$$
$$C_1 = -f(k_p) + f_3(k_p)\left(\frac{\lambda k_s^2}{2(3\lambda + 2\mu)\rho^3} + \frac{4}{\rho^5}\right),$$
$$A_2 = \frac{k_s^2}{\rho^3} f_3(k_p) + 2f(k_p),$$
$$B_2 = \frac{k_s^2}{\rho^3} f_3(k_s) + 2f(k_s),$$
$$C_2 = f_3(k_p)\frac{\lambda k_s^2}{2(3\lambda + 2\mu)\rho^3} \, ,$$
$$C_3 = f(k_p) + f_3(k_p)\frac{\nu}{(3\lambda + 2\mu)\rho^3} \, .$$

Proof. We conclude by resolving the linear system of equations (7.41), (7.61), (7.60) and (7.68) with respect to τ_{ij}. □

Note that the mean value theorem for the Lamé equation ($\nu = 0$) obtained in [6] is a special case of Theorem 7.5.

Corollary.

For the Lamé equation, ($\nu = 0$) the mean value relation reads

$$\tau_{ij}(0) = \frac{3}{8\pi(5+2\gamma)} \left\{ 10(1-\gamma) \int\limits_{\Omega} \tau_{ik} \frac{x_j x_k}{\rho^2}\, d\Omega \right.$$
$$\left. - \; 7\gamma\delta_{ij} \int\limits_{\Omega} \tau_{kl} \frac{x_k x_l}{\rho^2}\, d\Omega + 35\gamma \int\limits_{\Omega} \tau_{kl} \frac{x_k x_i x_j x_l}{\rho^4}\, d\Omega \right\} .$$

To show that, it is sufficient to take the limit in the last expression as $\nu \to 0$. □

Chapter 8

Applications to the Random Walk on Spheres method

8.1. Spherical mean as a mathematical expectation

The integral equations with a convergent Neumann series can be numerically solved by the Monte Carlo method using the Markov chains [10]. To make things clear we first illustrate the situation by a simple evaluation of the spherical mean (in $\mathbb{R}^3$) by the Monte Carlo method.

Let s be a random vector uniformly distributed on the sphere $S(x, r)$. Then by definition, the expectation of the random variable $u(s)$ is equal to the spherical mean $N_r u$:

$$\mathbb{E}u(s) = N_r u \equiv \frac{1}{4\pi r^2} \int\limits_{S(x,r)} u(y)\, dS(y) \ . \tag{8.1}$$

By the law of large numbers, the expectation is approximated by the arithmetic mean:

$$\mathbb{E}u(s) \approx U_N := \frac{1}{N} \sum_{i=1}^{N} u(\hat{s}^{(i)}), \tag{8.2}$$

where $\hat{s}^{(i)}$ are independent samples of the random vector s.

The statistical error of this approximation is measured by the quantity $\sigma/\sqrt{N}$, where σ^2 is the variance of $u(s)$. This means that with probability 0.997 the absolute difference between the arithmetic mean and the exact expectation is not larger than $3\sigma/\sqrt{N}$. It is supposed of course that $\sigma < \infty$ [40].

Thus we can formulate the Monte Carlo algorithm for calculation of the spherical mean over N samples:

P1. Put V:=0; i:=1;

P2. Sample a random vector s uniformly distributed on the sphere $S(x,r)$: $s = x + \zeta r$. Here $\zeta = (\zeta_1, \zeta_2, \zeta_3)$ is a unit isotropic vector which is simulated as follows:

2.1 $\zeta_1 = 1 - 2\alpha_0$;

2.2 $\gamma_1 = 1 - 2\alpha_1$; $\gamma_2 = 1 - 2\alpha_2$; $d = \gamma_1^2 + \gamma_2^2$;

2.3 If $d > 1$, then go to 2.2, otherwise

2.4 $\zeta_2 = \gamma_1\sqrt{(1-\zeta_1^2)/d}$; $\zeta_3 = \gamma_2\sqrt{(1-\zeta_1^2)/d}$;

{Here $\alpha_0, \alpha_1, \alpha_2$ are independent samples generated by the standard generator of pseudo-random numbers uniformly distributed on $[0,1]$}

P3. $V := V + u(s)$; if $i < N$ go to P2;

P4. The arithmetic mean (8.2) is then $U_N = V/N$.

8.2. Iterations of the spherical mean operator

In this section, let us use the following notation for the spherical mean $N_r^{x,y}[u]$ where the superscript indicates that the spherical mean is taken over the sphere $S(x,r)$, and y is the variable of integration over $S(x,r)$:

$$I_1 \equiv N_r^{x,y}[u] = \frac{1}{4\pi r^2} \int_{S(x,r)} u(y) dS(y) \ , \tag{8.3}$$

i.e., $y = x + rs$.

Suppose that it is desired to calculate the second iteration of the spherical mean

$$I_2 \equiv N_{r_1}^{x,y}[N_{r_2}^{y,z}[u]].$$

Introduce $Y_2(x)$, a set of Markov chains starting at the point x, with the first random state y_1 uniformly distributed on $S(x, r_1)$ and the second state y_2 uniformly distributed on $S(y_1, r_2)$, provided y_1 is fixed. On this Markov chain we define the random variable

$$\xi(x) = u(y_2). \tag{8.4}$$

It is obvious that

$$I_2 = \mathbb{E}\xi(x) = \mathbb{E}u(y_2),$$

where $\mathbb{E}$ denotes the expectation, that is, the average over all Markov trajectories $Y_2(x)$. Thus we can approximately evaluate

$$I_2 \approx \frac{1}{N} \sum_{j=1}^{N} u(y_2^{(j)}),$$

where $y_2^{(j)}$, $j = 1, 2 \ldots N$ are the second states of independently simulated Markov trajectories. Note that only the values of u at the last state y_2 are involved.

In the general case, to evaluate the k-th iteration

$$I_k \equiv N_{r_1}^{x,y_1} N_{r_2}^{y_1,y_2} \ldots N_{r_k}^{y_{k-1},y_k}[u] \tag{8.5}$$

we introduce a Markov chain $Y_k(x)$ starting at the point x and with the transitions $x \to y_1 \to y_2 \to \ldots \to y_k$ simulated according to the same scheme : the point y_i is uniformly distributed on the sphere $S(y_{i-1}, r_i)$, $i = 1, 2, \ldots, k$, $y_0 = x$. Then,

$$I_k = \mathbb{E}u(y_k).$$

8.3. The Random Walk on Spheres algorithm

8.3.1. The Random Walk on Spheres process for the Dirichlet problem

We show now that the spherical mean value relation and the Monte Carlo procedure of calculating the iterations of the spherical mean operator described above can be used to give a numerical method for solving the Dirichlet problem for the Laplace equation:

$$\Delta u(x) = 0, \quad x \in G, \qquad u|_\Gamma = \varphi \ .$$

We recall the definitions given in Chapter 2: $d(x)$ is the distance from the point x to the boundary of the domain, and Γ_ε is defined as the set

$$\Gamma_\varepsilon = \{x \in G : d(x) < \varepsilon\}.$$

Let us define the Random Walk on Spheres process in the domain G. It starts at the point x, and the states $y_1, y_2, \ldots$ are constructed as in the Markov chain $Y_k(x)$ described above, where $r_i = d(y_{i-1})$, and the process stops if after, say, N_ε steps, $y_{N_\varepsilon} \in \Gamma_\varepsilon$. We denote the Random Walk on Spheres process by Y_ε. This process is also called the ε-spherical process. It can be defined as a sequence of random points

$$y_i = y_{i-1} + d(y_{i-1})\omega_i,\ i = 1, 2, \ldots,\ y_0 = x,$$

where $\{\omega_i\}$ is a sequence of independent unit isotropic vectors. The last random state is y_{N_ε} such that $y_{N_\varepsilon} \in \Gamma_\varepsilon$ (see Fig. 8.1).

The main properties of the Random Walk on Spheres process Y_ε are well known (see, for example, [40]). The following property is important since it shows that the process rapidly converges to the boundary: for bounded domains of arbitrary dimension and practically arbitrary boundaries (for details, see [40]) the random process Y_ε with probability one converges to the boundary even if $\varepsilon = 0$ while the mean number of steps to reach Γ_ε is $O(\ln(\varepsilon))$ as $\varepsilon \to 0$. For unbounded domains, see [47].

Using the converse spherical mean value theorem (see Theorem 2.5) and the random estimators for the iterations I_k, it is possible to construct the random estimator for the Dirichlet problem. Indeed, since the Random Walk on Spheres process reaches the set Γ_ε with probability one in a finite number of steps, it follows from the converse spherical mean value theorem (see Section 2.1.2) that

$$u(x) = \mathbb{E}u(y_{N_\varepsilon})\ .$$

Choosing ε small enough, in practice one uses

$$u(x) \approx \mathbb{E}\varphi(\bar{y}_{N_\varepsilon})$$

where $\bar{y}_{N_\varepsilon}$ is the point of the boundary closest to the last state $y_{N_\varepsilon} \in \Gamma_\varepsilon$.

The convergence of the Random Walk on Spheres method follows from the estimation of the relevant integral operator generated by the spherical mean value relation. We present here the proof for the case of bounded domain $G \subset \mathbb{R}^3$ and domains with $d^* < \infty$, where d^* is a sphere of maximal radius which can be inscribed into the domain G.

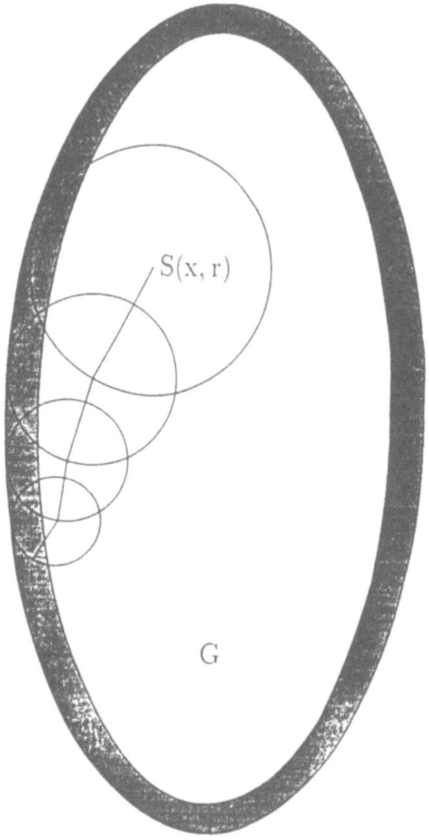

Figure 8.1. The random ε-spherical process starting at x, the center of the first sphere $S(x,r)$. In this illustration we see one trajectory which, after 4 steps, comes to Γ_ε.

More formally,

$$d^* = \sup_{x \in G} d(x),$$

where $d(x)$ is the radius of the sphere $S(x, d(x))$ centered at the point x inscribed into G; obviously, $d(x)$ equals the distance from the point $x \in G$ to Γ, the boundary of G. Let $P_\varepsilon(x)$ be the probability that the random point uniformly distributed on a sphere $S(x,r) \subset G$ hits the set Γ_ε. Simple arguments lead to the uniform estimation [8]

$$P_\varepsilon(x) \le \frac{\varepsilon^2}{4(d^*)^2} \ .$$

Using this estimation we get the convergence for the domains with $d^* < \infty$. To this end we first write down a formal integral equation generated by the spherical mean value relation (e.g., see [8]).

Let $\delta_x(y)$ be the density of the uniform mass on the sphere $S(x, d(x))$. We define the kernel function k_ε by

$$k_\varepsilon(x,y) := \begin{cases} \delta_x(y) & \text{if } x \in G \setminus \Gamma_\varepsilon \ , \\ 0 & \text{if } x \in \Gamma_\varepsilon \ , \end{cases} \tag{8.6}$$

and define the integral operator K_ε by

$$K_\varepsilon \psi(x) := \int_G k_\varepsilon(x,y)\psi(y)dy \tag{8.7}$$

for each $\psi(\cdot) \in C(G)$.

We now fix the boundary function φ. If G is unbounded we suppose that the solution to the Dirichlet problem tends to zero at infinity. Denote by u the solution to the Dirichlet problem corresponding to φ, and by $f_\varepsilon(x)$ the function

$$f_\varepsilon(x) = \begin{cases} 0 & \text{if } x \in G \setminus \Gamma_\varepsilon \ , \\ u(x) & \text{if } x \in \Gamma_\varepsilon \ . \end{cases}$$

Consider the integral equation

$$v(x) = K_\varepsilon v(x) + f_\varepsilon(x) \ . \tag{8.8}$$

Note that the mean value relation implies (8.8).

The relation between the equation (8.8) and the boundary value problem (2.1),(2.2) is given by the following proposition.

Proposition 8.1.

Suppose that the Dirichlet problem for the Laplace equation is uniquely solvable for any continuous function φ. Then for any $\varepsilon > 0$, the integral equation (8.8) has a unique solution given by the Neumann series

$$v(x) = f_\varepsilon(x) + \sum_{i=1}^{\infty} K_\varepsilon^i f_\varepsilon(x) \quad . \tag{8.9}$$

This solution coincides with the solution to the original Dirichlet problem.

Proof. Let ε be fixed. To show the convergence of the series

$$f_\varepsilon(x) + \sum_{i=1}^{N} K_\varepsilon^i f_\varepsilon(x) \quad , \tag{8.10}$$

it is sufficient to prove the existence of $0 < \lambda < 1$ such that, for any continuous and bounded function g,

$$\|K_\varepsilon^2 g\|_{L^\infty} < \lambda \|g\|_{L^\infty}$$

(this fact also implies the uniqueness of the solution to (8.9)). Let

$$\nu(\varepsilon) = \varepsilon^2/(4d^{*2}) \quad . \tag{8.11}$$

For $x \in G \setminus \Gamma_\varepsilon$ we have

$$\int_G k_\varepsilon(x,y) \int_G k_\varepsilon(y,y')dy'\, dy = \int_{G\setminus\Gamma_\varepsilon} \delta_x(y) \left(\int_G \delta_y(y')dy' \right) dy$$

$$= \int_{G\setminus\Gamma_\varepsilon} \delta_x(y)dy \le 1 - \nu(\varepsilon) < 1.$$

Let $v(x) := f_\varepsilon(x) + \sum_{i=1}^{\infty} K_\varepsilon^i f_\varepsilon(x)$. It is clear that v satisfies

$$v(x) = K_\varepsilon v(x) + f_\varepsilon(x) \quad .$$

If x belongs to Γ_ε, then, for any y, $k_\varepsilon(x,y) = 0$ and thus $v(x) = f_\varepsilon(x) = u(x)$. On the other hand, if x belongs to $G \setminus \Gamma_\varepsilon$, then the definition of k_ε implies that v satisfies the mean value relation at x with the sphere of radius $d(x)$. We conclude by using the converse mean value theorem. □

The proof presented does not work if $d^* = \infty$. We consider now two characteristic cases of domains with $d^* = \infty$: the first is the half-space

$\mathbb{R}^3_+$ and the second is the exterior of a ball. We show that in these two cases, the Random Walk on Spheres method converges.

Thus let us consider the ε-spherical process starting at a point x_0 in the half-space $\mathbb{R}^3_+$, R_0 being the distance from x_0 to the plane $\{z = 0\}$.

Theorem 8.1.

For $\lambda < \lambda_0 = \ln 2 - 1 - \ln\ln 2 \approx 0.0596$ the moment $\mathbb{E}\exp\lambda N_\varepsilon$ is finite and

$$\mathbb{E}\exp\lambda N_\varepsilon \le \frac{\exp\lambda}{2(1-\alpha)}\left(\frac{R_0}{\varepsilon}\right)^{\frac{\alpha}{\ln 2}-1} ,$$

where $\alpha < 1$ is the solution of the equation

$$\alpha\exp(-\alpha) = \frac{\ln 2}{2}\exp(\lambda) .$$

For the mean number of steps of the ε–spherical process the following estimation holds:

$$\mathbb{E}N_\varepsilon \le \frac{1}{2(1-\ln 2)} + \frac{\ln 2}{2(1-\ln 2)^3} + \frac{\ln(R_0/\varepsilon)}{2(1-\ln 2)^2} .$$

Proof. For arbitrary $\varepsilon > 0$ the exact representation of $P(N_\varepsilon = k)$, the probability that the number of steps is equal to k, $k > 1$, is

$$P(N_\varepsilon = k) = \frac{1}{2^k}\left(1-\frac{\varepsilon}{2R_0}\right)\frac{\varepsilon}{R_0}\int_\varepsilon^{2R_0} dR_1 \int_\varepsilon^{2R_1} dR_2$$

$$\dots \int_\varepsilon^{2R_{k-3}} dR_{k-2} \int_\varepsilon^{2R_{k-2}} dR_{k-1}\frac{1}{R_{k-1}}\prod_{j=1}^{k-2}\frac{1}{R_j}\left(1-\frac{\varepsilon}{2R_j}\right) .$$

Since $\varepsilon < 2R_j$ for all j, we get

$$\begin{aligned}\mathbb{E}\exp\lambda N_\varepsilon &\le \frac{\varepsilon\exp\lambda}{2R_0}\sum_{i=0}^{\infty}\left(\frac{\exp\lambda}{2}\right)^i\frac{\ln^i\{2^iR_0/\varepsilon\}}{i!}\\ &= \frac{\varepsilon\exp(\lambda)}{2R_0}\sum_{i=0}^{\infty}\left(\frac{\ln 2\exp\lambda}{2}\right)^i\frac{1}{i!}\Big[i+a\Big]^i\\ &=: f(a) ,\end{aligned}$$

where

$$a := \ln \left(\frac{R_0}{\varepsilon}\right)^{\frac{1}{\ln 2}} .$$

Direct calculations show that $f(a)$ satisfies the equation

$$\frac{\partial f}{\partial a} = \frac{\ln 2 \exp \lambda}{2} f(a+1) ,$$

whose solution has the form

$$f(a) = f(0) \exp(\alpha a) ,$$

where α is the unique solution on the interval $(0,1)$ to the equation

$$\alpha \exp(-\alpha) = \frac{\ln 2}{2} \exp \lambda .$$

The value $f(0)$ is easily found:

$$f(0) = \frac{\exp \lambda}{(1-\alpha)} \frac{\varepsilon}{2R_0} .$$

Thus,

$$\mathbb{E} \exp \lambda N_\varepsilon \leq \frac{\exp(\lambda)}{(1-\alpha)} \frac{\varepsilon}{2R_0} \exp \left\{ \alpha \ln \left(\frac{R_0}{\varepsilon}\right)^{\frac{1}{\ln 2}} \right\} = \frac{\exp \lambda}{2(1-\alpha)} \left(\frac{R_0}{\varepsilon}\right)^{\frac{\alpha}{\ln 2}-1} .$$

Note that $f(0) < \infty$ if $\ln 2 \exp(\lambda) < 2/e$, i.e., when $\lambda < \lambda_0 = \ln 2 - 1 - \ln \ln 2 \approx 0.0596$.

The solution to $\alpha \exp(-\alpha) = \ln 2 \exp(\lambda/2)$ on $(0,1)$ can be found numerically, e.g., for $\lambda = 0.05$ we get $\alpha = 0.86$ and

$$\mathbb{E} \exp \lambda N_\varepsilon \leq 3.754 \left(\frac{R_0}{\varepsilon}\right)^{0.24} .$$

For the mean number of steps we get

$$\mathbb{E} N_\varepsilon \leq \frac{\varepsilon}{2R_0} \sum_{i=0}^{\infty} \left(\frac{\ln 2}{2}\right)^i (i+1) \frac{1}{i!} (i+a)^i .$$

Let

$$F(\beta) := \frac{\varepsilon}{2R_0} \sum_{i=0}^{\infty} \left(\frac{\ln 2}{2}\right)^i \beta^{i+1} \frac{1}{i!} (i+a)^i \quad .$$

Then,

$$\mathbb{E} N_\varepsilon \le \frac{\partial F}{\partial \beta}|_{\beta=1} \quad .$$

But

$$F(\beta) = \frac{\beta}{2(1-\alpha)} \left(\frac{R_0}{\varepsilon}\right)^{\frac{1}{\ln 2}-1} \quad ,$$

where $\alpha \exp(-\alpha) = \frac{\beta \ln 2}{2}$. The derivative gives:

$$\frac{\partial F}{\partial \beta}|_{\beta=1} = \frac{1}{2(1-\ln 2)} + \frac{\ln 2}{2(1-\ln 2)^3} + \frac{\ln(R_0/\varepsilon)}{2(1-\ln 2)^2} \quad .$$

The theorem is proved. □

The case $d^* = \infty$ is a challenging problem in two respects: first, the behavior of the Random Walk on Spheres process is more complicated since there is a possibility that the process goes to infinity. Second, the standard approach [40] for constructing the random estimators for the Dirichlet problem in the bounded domain fails.

It should be noted that, if the problem is solved in $G_1 = \mathbb{R}^3 \setminus G$, where G is a finite domain, then the Random Walk on Spheres algorithm can be constructed as follows (e.g. see [9]). We imbed the domain G into a sphere $S(y, R_0)$: $G \subset S(y, R_0)$; construct the Random Walk on Spheres process in the domain G_1; outside of $S(y, R_0)$ use the transition to the surface of the sphere $S(y, R_0)$ according to the probability distribution generated by the Poisson kernel of the exterior Dirichlet problem for a ball. Actually, this is a special case of the method of iterations over subdomains described in [40]. This approach is restricted however to the case of bounded G. It reduces the problem to the standard Random Walk on Spheres process in bounded domains and says nothing about the behavior of this process at infinity.

Although we study in detail a simple case when G is a sphere, it is believed that it provides the essential properties of the Random Walk on Spheres process in unbounded domains and gives an idea of how to treat more general cases. In practical problems one always tries to avoid simulation of long trajectories by different techniques, for instance, using the explicit form of the Green functions as described above. However it is of fundamental importance to investigate the Random Walk on Spheres process in the general case. In particular, the estimation of the exponential moment provides a probabilistic representation in the general case. With this representation, it is then possible to combine different strategies to improve the efficiency of the numerical algorithms (see [41] and [45]).

Let us consider the unbounded domain $G_1 \subset \mathbb{R}^3$ which is the exterior to the sphere $S(z,r)$ of radius r centered at a point $z \in G_1$. We define also an ε-layer

$$\Gamma_\varepsilon = \left\{x \in G_1 : r \leq |x - z| \leq r + \varepsilon\right\}.$$

The Random Walk on Spheres process in G_1 starting at a point $x \in G_1$ is defined as the Markov process $\{\xi_0, \xi_1, \ldots, \xi_n, \ldots\}$ such that

$$\xi_0 = x,\ \xi_{k+1} = \xi_k + d(\xi_k)\omega_k, \quad k = 0, 1, \ldots, \tag{8.12}$$

where $d(y)$ is the distance from the point y to the sphere $S(z,r)$, and $\omega_0, \omega_1, \ldots,$ is a set of independent unit isotropic random vectors in $\mathbb{R}^3$.

So defined, the Random Walk on Spheres process consists of an infinite number of states $\{\xi_0, \xi_1, \ldots, \xi_k, \ldots\}$. We define the ε-spherical process $\Xi_\varepsilon(x)$ as the Random Walk on Spheres process starting at x and which terminates if after, say, N_ε steps $\xi_{N_\varepsilon} \in \Gamma_\varepsilon$.

The number of steps, N_ε, is random, and it is of interest to derive different statistical characteristics of it, such as the mean number EN_ε, the exponential moment $E\exp(\lambda N_\varepsilon)$, etc.

Theorem 8.2.

Let P_k be the probability that the ε-spherical process $\Xi_\varepsilon(x)$ terminates after k steps. Then

$$P_{k+1} \leq \frac{\varepsilon + 2r}{2(r + d(x))}\frac{\varepsilon}{2d(x)}\frac{\ln^k\{2^k d(x)/\varepsilon\}}{2^k k!}, \quad k = 0, 1, \ldots, \tag{8.13}$$

and

$$E\exp\{\lambda N_\varepsilon\} \leq \frac{\varepsilon + 2r}{2(r + d(x))} \frac{\exp(\lambda)}{2(1-\alpha)} \left(\frac{d(x)}{\varepsilon}\right)^{\frac{\alpha}{\ln 2}-1} \tag{8.14}$$

for $\lambda < \lambda_0 = \ln 2 - 1 - \ln\ln 2 \approx 0.0596$. *Here* α *is the unique (on the interval* $(0,1)$*) solution to the equation*

$$\alpha\exp(-\alpha) = \frac{\ln 2}{2}\exp(\lambda).$$

Proof. Let $R_0 = d(x)$. Then

$$P(N_\varepsilon = 1) = \frac{2\pi h_0 R_0}{4\pi R_0^2} = \frac{h_0}{2R_0}, \tag{8.15}$$

where h_0 is defined from the obvious equation

$$(r+\varepsilon)^2 - (r+h_0)^2 = R_0^2 - (R_0 - h_0)^2,$$

namely,

$$h_0 = \frac{\varepsilon^2 + 2r\varepsilon}{2(R_0 + r)}.$$

The probability that the ε-process has exactly two steps is

$$P(N_\varepsilon = 2) = \left(1 - \frac{h_0}{2R_0}\right)\frac{1}{2R_0}\int_{h_0}^{2R_0} \frac{h_1}{2R_1}\,dx_0, \tag{8.16}$$

where the integration is carried out with respect to the measure dx_0, x_0 varying between 0 and $2R_0$, being the projection of the center of the second sphere on the ray z, z_0; h_1 is found in the second sphere as h_0 was found in the first one:

$$h_1 = \frac{\varepsilon^2 + 2r\varepsilon}{2(R_1 + r)},$$

where R_1 is the radius of the second sphere in the Random Walk on Spheres process. Note that

$$(r + R_1)^2 - (r + x_0)^2 = R_0^2 - (R_0 - x_0)^2;$$

hence,

$$x_0 = \frac{R_1(R_1 + 2r)}{2(R_0 + r)}.$$

Using this relation we rewrite (8.16)

$$P(N_\varepsilon = 2) = \left(1 - \frac{h_0}{2R_0}\right)\frac{1}{2R_0}\frac{\varepsilon^2 + 2r\varepsilon}{2(R_0 + r)}\int\limits_{\varepsilon}^{2R_0}\frac{dR_1}{2R_1}. \tag{8.17}$$

On the kth step we get

$$P(N_\varepsilon = k) = \left(1 - \frac{h_0}{2R_0}\right)\frac{1}{2R_0}\int\limits_{h_0}^{2R_0}\left(1 - \frac{h_1}{2R_1}\right)\frac{dx_0}{2R_1}\int\limits_{h_1}^{2R_1}\left(1 - \frac{h_2}{2R_2}\right)\frac{dx_1}{2R_2}$$

$$\dots\int\limits_{h_{k-2}}^{2R_{k-2}}\left(1 - \frac{h_{k-1}}{2R_{k-1}}\right)\frac{dx_{k-2}}{2R_{k-1}}\int\limits_{h_{k-1}}^{2R_{k-1}}\frac{h_k}{2R_k}\,dx_{k-1}. \tag{8.18}$$

Using the relations

$$h_i = \frac{\varepsilon^2 + 2r\varepsilon}{2(R_i + r)},$$

$$x_{i-1} = \frac{R_i(2r + R_i)}{2(R_{i-1} + r)},$$

we successively get from (8.18)

$$P(N_\varepsilon = k) = \left(1 - \frac{h_0}{2R_0}\right)\frac{\varepsilon^2 + 2r\varepsilon}{2R_0(R_0 + r)}\int\limits_{\varepsilon}^{2R_0}\left(1 - \frac{h_1}{2R_1}\right)\frac{dR_1}{2R_1}$$

$$\dots\int\limits_{\varepsilon}^{2R_{k-3}}\left(1 - \frac{h_{k-2}}{2R_{k-2}}\right)\frac{dR_{k-2}}{2R_{k-2}}\int\limits_{\varepsilon}^{2R_{k-2}}\frac{dR_{k-1}}{2R_{k-1}},$$

or,

$$P(N_\varepsilon = k) = \left(1 - \frac{h_0}{2R_0}\right) \frac{\varepsilon^2 + 2r\varepsilon}{2(R_0 + r)R_0} \left(\frac{1}{2}\right)^k \int_\varepsilon^{2R_0} dR_1 \int_\varepsilon^{2R_1} dR_2$$
$$\cdots \int_\varepsilon^{2R_{k-3}} dR_{k-2} \int_\varepsilon^{2R_{k-2}} \frac{dR_{k-1}}{R_{k-1}} \prod_{j=1}^{k-2} \frac{1}{R_j}\left(1 - \frac{h_j}{2R_j}\right). \tag{8.19}$$

Note that the expression (8.19) differs from the representation of $P(N_\varepsilon = k)$ obtained in the Theorem 8.1 for the half space $\mathbb{R}^3_+$ only by the factor

$$\frac{\varepsilon + 2r}{2(R_0 + r)} \cdot \tag{8.20}$$

Thus we conclude by using the results of the Theorem 8.1. □

Note that, from the results given in the above theorem, it is seen that there is a portion of trajectories ($\approx 1 - (\varepsilon + 2r)/2(r + R_0)$) which go to infinity. Since, however, $u(x) \to 0$ at infinity, the conditional expectation coincides with the total expectation. From these estimations it is possible to obtain a practical scheme of implementation of the algorithm for solving exterior boundary value problems for domains whose boundaries have a complicated shape. For instance, to reach the probabilistic error of order ε, it is necessary to ignore the trajectories which are out of the sphere of radius

$$R_\infty \sim r + \frac{r(\rho - r)}{\rho\varepsilon}$$

where ρ is the distance from the point where the solution is calculated to the center of the sphere $S(z, r)$. Then

$$u(x) \approx \langle \varphi(\bar{y}_{N_\varepsilon}) \rangle$$

where $\bar{y}_{N_\varepsilon} \in \Gamma$ is the point closest to the final state $y_{N_\varepsilon} \in \Gamma_\varepsilon$.

It is interesting to note that similar arguments in a two-dimensional case lead to the estimation $R_\infty \sim r(\rho/r)^{1/\varepsilon}$ which implies that to reach the desired probabilistic accuracy ε it is necessary to simulate much longer trajectories compared to the In $3D$ case.

Remark 8.1.

The same scheme works for equations for which the converse spherical mean value theorem can be proved and, in addition, the convergence of the iterative procedure for the relevant integral equation can be shown. For example, let us consider the Dirichlet problem for the diffusion equation ($\lambda \geq 0$):

$$\Delta u(x) - \lambda u(x) = 0, \quad u|_\Gamma = \varphi.$$

The random estimator which can be used to calculate the approximate solution is

$$\xi = \varphi(\bar{y}_{N_\epsilon}) \prod_{i=0}^{N_\epsilon - 1} \frac{d(y_i)\sqrt{\lambda}}{\sinh[d(y_i)\sqrt{\lambda}]}.$$

8.3.2. Inhomogeneous case

Let us consider the Dirichlet problem for the Poisson equation:

$$\Delta u(x) = f(x), \quad x \in G\ , \qquad u|_\Gamma = 0\ .$$

We construct now a Random Walk on Spheres process which solves this problem. We start by the integral representation

$$u(x_0) = \int_G u_\delta(x, x_0) f(x) dx\ , \tag{8.21}$$

where u_δ is the Green function, i.e.,

$$\Delta u_\delta(x, x_0) = \delta(x - x_0), \quad u_\delta|_\Gamma = 0\ .$$

We seek the Green function u_δ in the form

$$u_\delta(x, x_0) = \mathcal{E}(x, x_0) + W(x, x_0)\ ,$$

where $\mathcal{E}(x, x_0)$ is the fundamental solution to the Laplace equation, and hence W is the solution to

$$\Delta W(x, x_0) = 0, \quad W|_{x \in \Gamma} = -\mathcal{E}(x, x_0)|_{x \in \Gamma}\ . \tag{8.22}$$

Thus we see that the problem is reduced to calculating the integral functional (8.21) of the solution to the Dirichlet problem for the Laplace equation (8.22). The integral (8.21) can be represented as the expectation

$$u(x_0) = \langle \eta(y, x_0) \rangle \ ,$$

where the random estimator $\eta(y, x_0)$ is constructed as follows:

$$\eta(y, x_0) = u_\delta(y, x_0) f(y)/\pi(y) \ ,$$

and y is a random point sampled in G from an appropriate probability density function $\pi(y)$. Since for the Green function we also can construct a random estimator, say, $\xi(x_0, y)$ so that $u_\delta(x_0, y) = \mathbb{E}_y \xi(x_0, y)$, we come to the representation

$$u(x_0) = \langle \mathbb{E}_y \xi(x_0, y) \rangle \tag{8.23}$$

so that

$$\eta(y, x_0) = \frac{f(y)}{\pi(y)} \{ \mathcal{E}(y, x_0) + \zeta(y, x_0) \} \ , \tag{8.24}$$

since $\xi(x_0, y) = \mathcal{E}(y, x_0) + \zeta(y, x_0)$ where $\zeta(y, x_0)$ is the standard estimator of the Random Walk on Spheres algorithm for the Dirichlet problem (8.22).

Thus the double expectation (8.23) can be evaluated as follows.

A point y is chosen at random in G according to the density $\pi(y)$ and a δ-source is placed at y. Then the Random Walk on Spheres process starting from x_0 is constructed and the estimator (8.24) is calculated. So along one trajectory a contribution from the δ-source at y is calculated. Next a new random point y_1 is sampled according to the same density and the contribution from the δ-source at $y = y_1$ is calculated, etc. The final result is obtained by averaging the contributions over a sufficiently large number of points $y, y_1, \dots, y_p$.

The algorithm described gives the solution only at one fixed point x_0. However it is possible to construct a modification which gives the solution at a set of points. This modification is based on the symmetry of the Green function: $u_\delta(x, x_0) = u_\delta(x_0, x)$. Let us describe this algorithm which gives the solution at a set of points $x_i, \quad i = 1, \dots, m$.

The δ-sources are located at the points $x_i, \quad i = 1, \dots, m$ and fixed for all trajectories. A random point y_1 is sampled as above, from the density $\pi(y)$. The Random Walk on Spheres process starting from this point, y_1, is simulated, and contributions from the δ-sources are calculated, i.e., the estimators $\eta(x_i, y_1)$ are calculated for $i = 1, \dots, m$. Next a new random point, y_2, is sampled, the Random Walk on Spheres process starting from y_2 is constructed, and new contributions $\eta(x_i, y_1)$ are calculated for $i = 1, \ldots, m$. Again, the result is obtained by averaging the contributions over a sufficiently large number of samples $y_1, \ldots, y_p$.

Details of these algorithms can be found in [40].

Note that it was assumed that $\varphi = u|_\Gamma = 0$. In the general case, when $\varphi \neq 0$, a different approach can be applied. In this case we need the volume mean value relation in a ball, i.e., the Green formula for a ball:

$$u(x) = \int\limits_{S(x,r)} \frac{\partial u_\delta(x,y)}{\partial n} \varphi(y)\, dS(y) + \int\limits_{B(x,r)} u_\delta(x,x') f(x')\, dV(x') \ .$$

In [40], a modification of the Random Walk on Spheres based on this Green formula is described. It appears that along the random point on each sphere, it is necessary to sample a random point in the ball. For details, see [40] and [8].

8.4. Biharmonic equation

The vector case is treated in the same way (see [39] and [42]). Let us consider, for example, the boundary value problem (3.9),(3.7), i.e. the biharmonic equation

$$\Delta^2 u(x) = 0 \quad , \quad x \in G \subset \mathbb{R}^m$$

with boundary conditions

$$u\Big|_\Gamma = g_0, \quad \left\{\Delta u - \frac{1-\sigma}{\rho}\frac{\partial u}{\partial n}\right\}\Bigg|_\Gamma = g_1 \ .$$

To carry out an iterative procedure, we write down the integral equation, as we have done in Chapter 2. We can rewrite the mean value relation in the form of the following system of integral equations for the vector $\mathbf{v} = (v_1, v_2)^T$:

$$\mathbf{v} = K_\varepsilon \mathbf{v} + \mathbf{f}_\varepsilon \ , \tag{8.25}$$

where the matrix-kernel $k_\varepsilon(x,y)$ and the vector $\mathbf{f}_\varepsilon$ have the form

$$k_\varepsilon(x,y) = \begin{cases} \begin{pmatrix} \delta_x(y) & -\frac{r^2}{2m}\delta_x(y) \\ 0 & \delta_x(y) \end{pmatrix} & , \quad x \in G \setminus \Gamma_\varepsilon \\ 0 & , \quad x \in \Gamma_\varepsilon \end{cases}$$

and

$$\mathbf{f}_\varepsilon(x) = \begin{cases} \begin{pmatrix} u_1(x) \\ u_2(x) \end{pmatrix} & , \quad x \in \Gamma_\varepsilon \\ 0 & , \quad x \in G \setminus \Gamma_\varepsilon \ . \end{cases}$$

The solution to the preceding integral equation is given by

$$\mathbf{v}(x) = \sum_{n=0}^{\infty} K_\varepsilon^n \mathbf{f}_\varepsilon(x) \ ,$$

where K_ε is the matrix-integral operator corresponding to the kernel $k_\varepsilon(x,y)$:

Proposition 8.2.

For arbitrary $d^* > \varepsilon > 0$, the function v is well defined.

Proof. The following estimate holds:

$$\|K_\varepsilon^n\|_{L^\infty} \leq [1 - \nu(\varepsilon)]^{n-1} \left(1 + \frac{n(d^*)^2}{2m}\right) \ . \tag{8.26}$$

Hence, ε being fixed, one can choose n large enough so that the above series converges. □

As in Chapter 2, we can prove that the solution to the above integral equation coincides with the solution of the boundary value problem (3.9), (3.7).

The kernel of the system of integral equations is the matrix

$$K(r) = \begin{pmatrix} 1 & \frac{-r^2}{2m} \\ 0 & 1 \end{pmatrix} .$$

Consequently, $\xi = (\xi_1, \xi_2)$, and the vector random estimator for $\mathbf{v} = (u, \Delta u)^T$ is

$$\xi(x) = \left\{ \prod_{i=1}^{N_\varepsilon} K(d(y_{i-1})) \right\} \mathbf{v}(y_{N_\varepsilon}),$$

where $y_0 = x$.

This means that

$$\mathbf{v}(x) = \mathbb{E}\xi.$$

Again, the ε-biased estimator has the form

$$\hat{\xi}(x) = \prod_{i=1}^{N_\varepsilon} \left[K(d(y_{i-1})) \right] \mathbf{g}(\bar{y}_{N_\varepsilon}),$$

where $\mathbf{g} = (g_0, g_1)^T$.

Note that since the matrix K is triangular, the random estimators ξ_2 and $\hat{\xi}_2$ for the solution to (3.9), (3.7) can be rewritten in the form

$$\xi_2 = u(X_{N_\varepsilon}) - \frac{\Delta u(X_{N_\varepsilon})}{2m} \sum_{k=0}^{N_\varepsilon - 1} d^2(X_k) \ ,$$

$$\hat{\xi}_2 = g_0(\bar{X}_{N_\varepsilon}) - \frac{g_1(\bar{X}_{N_\varepsilon})}{2m} \sum_{k=0}^{N_\varepsilon - 1} d^2(X_k) \ .$$

8.5. Random Walk on Spheres method for the Lamé equation

8.5.1. Naive generalization

In Chapter 4 we treated triangular systems of elliptic equations generating triangular systems of integral equations which in turn leads to a simple analysis of the convergence of the iterative process.

Let us consider a system in $G \subset \mathbb{R}^n$ where all the equations are symmetrically entered :

$$\mu \Delta u_1(x) + (\lambda + \mu) \tfrac{\partial}{\partial x_1} (\operatorname{div} \mathbf{u}) = 0 \ ,$$
$$\vdots$$
$$\mu \Delta u_n(x) + (\lambda + \mu) \tfrac{\partial}{\partial x_n} (\operatorname{div} \mathbf{u}) = 0 \ ,$$

or, using the summation convention (i fixed, summation over the repeated index j) :

$$u_{i,jj} + \alpha u_{j,ji} = 0 \ , \quad x \in G \subset \mathbb{R}^n \ , \tag{8.27}$$

where $i, j = 1, 2, \dots, n$, $\alpha = (\lambda + \mu)/\mu$, $\lambda, \mu > 0$ are constants.

This is the equilibrium equation for the displacements $u_i(x)$, $i = 1, 2, \dots, n$, the body forces being absent, and it is known as the Lamé equation; λ, μ are the Lamé constants of the elastic material.

We deal in this section with the first boundary value problem for (8.27) with the boundary conditions :

$$\mathbf{u}|_\Gamma = \varphi = (\varphi_1, \dots, \varphi_n)^T \ . \tag{8.28}$$

We rewrite the mean value relation (5.7) (see Theorem 5.1 in Section 5.1.1) in the vector form

$$\mathbf{u} = p\mathcal{N}_r\mathbf{u} + q\mathcal{S}_r\mathbf{u} \ , \tag{8.29}$$

where

$$p = a = \frac{n(2-\alpha)}{2(n+\alpha)} \geq 0 \ , \quad q = \beta = \frac{\alpha(n+2)}{2(n+\alpha)} \geq 0 \ ;$$

hence $p + q = 1$, and $\mathcal{N}_r$ is the matrix integral operator

$$\mathcal{N}_r\mathbf{u} = \begin{pmatrix} N_r u_1 & 0 & \cdots & 0 \\ 0 & N_r u_2 & \cdots & 0 \\ \vdots & & \ddots & \vdots \\ 0 & \cdots & & N_r u_n \end{pmatrix}$$

and

$$(\mathcal{S}_r\mathbf{u})_i(x) = \frac{n}{\omega_m}\int_\Omega s_i s_j u_j(x + rs)\,d\Omega(s) \ , \quad i, j = 1, 2, \dots, n \ . \tag{8.30}$$

A naive vector Monte Carlo estimator can be constructed as a generalization of the standard isotropic Random Walk on Spheres method. We introduce the unbiased vector estimator

$$\xi_\varepsilon = \prod_{i=0}^{N_\varepsilon - 1} (pI + qS_i)\mathbf{u}(X_{N_\varepsilon}) \ , \tag{8.31}$$

and the ε-biased random estimator

$$\bar{\xi} = \prod_{i=0}^{N_\epsilon - 1} (pI + qS_i)\varphi(\bar{X}_{N_\epsilon}) , \qquad (8.32)$$

where I is the $n \times n$ identity matrix, and S_i is the matrix-kernel of the operator (8.30) in i-th sphere:

$$S_k = n \begin{pmatrix} s_1^2 & s_1 s_2 & \cdots & s_1 s_n \\ s_2 s_1 & s_2^2 & \cdots & s_2 s_n \\ \vdots & & \ddots & \vdots \\ s_m s_1 & \cdots & & s_n^2 \end{pmatrix} . \qquad (8.33)$$

Thus $s_1, s_2, \ldots, s_n$ are the components of the unit isotropic vector in i-th sphere of the Random Walk on Spheres process.

From the integral formulation of the Lamé equation given in Chapter 5 it follows that

$$\mathbf{u}(x) = \mathbb{E}_x \hat{\xi}_\varepsilon . \qquad (8.34)$$

However, the estimators (8.31),(8.32) cannot be used if ε is small enough. Indeed, the variance of $\hat{\xi}(x)$ is exponentially increasing: $\mathbb{E}(\hat{\xi}_\varepsilon)^2 \cong n^{\bar{N}}$, where $\bar{N}$ is the number of steps of the ε-spherical process.

8.5.2. A modification of the algorithm

It is known that if ε is small, then after a certain number of steps the trajectory is concentrated near the boundary, r_i being of order of ε (see, e.g., [40] and [28]). This gives the motivation to construct the following modification of the algorithm.

Let the constants C_{ij} and D_{ij} be such that

$$\sup_{\Gamma_\epsilon} \left| \frac{\partial^2 u_i}{\partial x_i \partial x_j} \right| = C_{ij} , \quad \sup_{\Gamma_\epsilon} \left| \frac{\partial^2 \Delta u_i}{\partial x_i \partial x_j} \right| = D_{ij} , \quad (i \neq j)$$

and set

$$C_i = \sum_{i \neq j} C_{ij} , \quad D_i = \sum_{i \neq j} D_{ij} .$$

Let us consider the sum of the non-diagonal terms of the operator (8.29)

$$m_i = \frac{1}{4\pi}\int_\Omega s_i\, s_j\, u_j(s)\, d\Omega(s) \ , \quad (j \neq i) \ .$$

From the expansions of Lemma 4.1 we get in the three-dimensional case :

$$m_i = \frac{r^2}{15}\frac{\partial}{\partial x_i}u_{j,j} + \frac{r^4}{210}\frac{\partial}{\partial x_i}\Delta u_{j,j} \ ,$$

where the summation is taken over $j \neq i$.

In the 3D case, we rewrite (8.29) in the form :

$$u_i(x) = p(\mathcal{N}_r\mathbf{u})_i(x) + 3q\int_\Omega s_i^2\, u_i(x+rs)\, d\Omega(s) + m_i \ .$$

Thus, in Γ_ε we can approximate the local integral equation (8.29) by

$$\hat{u}_i(x) = p(\mathcal{N}_r\hat{\mathbf{u}})_i(x) + 3q\int_\Omega s_i^2\, \hat{u}_i(x+rs)d\Omega(s) + \frac{\hat{d}^2(x)}{15}\, C_i + \frac{\hat{d}^4(x)}{210}\, D_i \ , \tag{8.35}$$

where

$$\hat{d}(x) = \begin{cases} d(x) \ , & x \in \Gamma_\varepsilon \ , \\ 0 \ , & x \notin \Gamma_\varepsilon \ . \end{cases}$$

Define a diagonal matrix-operator $\mathcal{C}$ by

$$(\mathcal{C}\mathbf{v})_i(x) = \frac{3}{4\pi}\int_\Omega s_i^2\, v_i(x+rs)\, d\Omega(s) \ , \quad S(x,r) \subset G \ ,$$

(no summation over i).

By $p+q=1$, we obtain that the Neumann series for (8.35) absolutely converges for an arbitrary small ε, since the norm of the second iteration of the integral operator in (8.35) has the following estimate :

$$\|(p\mathcal{N}_r + q\mathcal{C})^2\|_{L^\infty(G)} < 1 - \nu(\varepsilon) < 1 \ .$$

From this we can estimate the error corresponding to (8.35) :

$$\sup_{i,\,\Gamma_\varepsilon} |u_i - \hat{u}_i| \approx \max_i\,(C_i,\ D_i) \sup_{x\in\Gamma_\varepsilon} \sum_{k=1}^{\infty} (p\mathcal{N}_r + q\mathcal{C})^k\, \hat{d}^2(x) \approx \max_i\,(C_i, D_i)\frac{\varepsilon^2}{\nu(\varepsilon)}$$

as $\varepsilon \to 0$.

Thus for smooth boundaries for which $\nu(\varepsilon) \approx \varepsilon$, we conclude that the error is of order ε.

Remark 8.2.

Note that the mean value relation

$$u_i(x) = p(\mathcal{N}_r\mathbf{u})_i(x) + q(\mathcal{C}\mathbf{u})_i(x) \ , \quad S(x,r) \subset G$$

corresponds to the case when $u_{i,ji} = 0$. This implies that we have actually three independent diffusion equations

$$\Delta u_i + \alpha \frac{\partial^2 u_i}{\partial x_i^2} = 0 \ .$$

Near the boundary, where r_i are small, the regular solutions to these diffusion equations have the mean value property

$$u_i(x) = p(\mathcal{N}_r\mathbf{u})_i(x) + q(\mathcal{C}\mathbf{u})_i(x) + O(r^4) \ , \quad S(x,r) \subset G \ .$$

The approximation (8.35) corresponds to the approximation of the Lamé equation near the boundary by three independent diffusion equations

$$u_{i,jj} + \alpha u_{i,ii} = 0 \ .$$

Thus, the modified algorithm is the following.

First, choose $\delta = t\varepsilon$ where $t > 1$ is a constant. Evaluate (8.31), until $X_m \in \Gamma_\delta$. After that, use the operator (8.35), omitting the right-hand side. The estimator is then

$$\tilde{\xi}_\varepsilon = \varphi(\bar{X}_{N_\varepsilon}) \prod_{i=0}^{N_\varepsilon - 1} (pI + qC_i) \ .$$

Note that the operator $\mathcal{C}$ has a probabilistic interpretation: we make with the probability p the standard isotropic step of the ε-spherical process, and with probability $q = 1-p$ the random point X_{n+1} is sampled on $S(X_n, d(X_n))$ with the density ms_i^2.

Of course, this procedure is efficient if the constants C_{ij} and D_{ij} are not too large. Thus, in the general case, it is not possible to improve the algorithm by this approach.

8.5.3. Non-isotropic Random Walk on Spheres

We now construct a different estimator which uses the specific structure of the mean value relation. The main idea is to find a distribution on the sphere which is in some sense generated by the matrix-kernel of the mean value relation. It appears that the points on the sphere should then have the "s_i"-distribution introduced above.

We begin with the case $p = 0,\ q = 1$. It is then necessary to evaluate the iterations of the operator (8.30) along the spheres $S(X_i, d(X_i))$ where we have to define the points X_i.

Let $X_0 = x, X_1, \ldots, X_m$ be points on the spheres

$$S(X_0, d(X_0)), S(X_1, d(X_1)), \ldots, S(X_m, d(X_m)) ,$$

respectively. We have to evaluate the products

$$\prod_{i=1}^{m} S_i ,$$

where the matrix S_i is defined in (8.33).

Let $s^{(i)},\ \ i = 1, \ldots, m$ be the unit vector which specifies the corresponding point X_i on the i-th sphere. We use the following representation (which is easily shown by induction):

$$\prod_{i=1}^{m} S_i = n^m (s^{(1)} \cdot s^{(2)}) \ldots (s^{(m-1)} \cdot s^{(m)})\ [s^{(1)} \otimes s^{(m)}], \tag{8.36}$$

where

$$[s^{(1)} \otimes s^{(m)}] = \begin{pmatrix} s_1^{(1)} s_1^{(m)} & s_1^{(1)} s_2^{(m)} & \ldots & s_1^{(1)} s_n^{(m)} \\ s_2^{(1)} s_1^{(m)} & s_2^{(1)} s_2^{(m)} & \ldots & s_2^{(1)} s_n^{(m)} \\ \vdots & & \ddots & \vdots \\ s_n^{(1)} s_1^{(m)} & \ldots & & s_n^{(1)} s_n^{(m)} \end{pmatrix} .$$

To be more specific, let us first consider the two dimensional case. The first m iterations give, using (8.36):

$$u_1(x) = \frac{2^m}{\pi^m} \int_{\Omega^m} R(\psi_1, \ldots, \psi_m) \frac{\cos \psi_1}{2} (\mathbf{u}, s^{(m)}) d\psi_1 \ldots d\psi_m \tag{8.37}$$

and

$$u_2(x) = \frac{2^m}{\pi^m} \int_{\Omega^m} R(\psi_1, \ldots, \psi_m) \frac{\sin \psi_1}{2} (\mathbf{u}, s^{(m)}) d\psi_1 \ldots d\psi_m, \tag{8.38}$$

where

$$R(\psi_1, \ldots, \psi_m) = \frac{1}{2^{m-1}} \prod_{j=1}^{m-1} \cos(\psi_{j+1} - \psi_j),$$

$$(\mathbf{u}, s^{(m)}) = u_1 \cos(\psi_m) + u_2 \sin(\psi_m).$$

We now give a probabilistic interpretation to (8.37).

Let us define a distribution density on the half-circle $\{s^{(1)} > 0\}$ by

$$p_{1+}(s) = \frac{\cos \psi}{2}, \quad -\frac{\pi}{2} \leq \psi \leq \frac{\pi}{2}. \tag{8.39}$$

On the second circle we sample the random direction from the conditional density

$$p_{2+}(s) = \frac{\cos(\psi_2 - \psi_1)}{2}, \quad -\frac{\pi}{2} + \psi_1 \leq \psi_2 \leq \frac{\pi}{2} + \psi_1, \tag{8.40}$$

under the condition that ψ_1 was sampled in the first circle from the density (8.39), etc., including the direction $s^{(m)}$ in the last circle.

The densities p_{i-} , $i = 1, \ldots, m$ in the half-circles $\{s^{(1)} \leq 0\}$ are constructed analogously.

In the i-th circle, we sample the random direction with probability 1/2 from $p_{i+}(s)$, and with probability 1/2 from $p_{i-}(s)$.

Let N_1 be the number of steps until $X_{N_1} \in \Gamma_\varepsilon$, and let κ_{N_1} be the number of samples that occurred in $\{s^{(1)} \leq 0\}$. Then (8.37) can be rewritten in the form of the expectation taken over the ensemble of directions $\{s^{(i)}\}_{i=1}^{N_1}$:

$$u_1(x) = \mathbb{E}\hat{\xi}_1(x), \tag{8.41}$$

where

$$\hat{\xi}_1(x) = \left(\frac{4}{\pi}\right)^{N_1} (-1)^{\kappa_{N_1}} [u_1(X_{N_1}) \cos \psi_{N_1} + u_2(X_{N_1}) \sin \psi_{N_1}]. \tag{8.42}$$

The estimator for the second component is constructed analogously:

$$\hat{\xi}_2(x) = \left(\frac{4}{\pi}\right)^{N_2} (-1)^{\kappa_{N_2}} [u_1(y_{N_2}) \cos \psi_{N_2} + u_2(y_{N_2}) \sin \psi_{N_2}], \tag{8.43}$$

where the process $\{y_i\}_{i=1}^{N_2}$ is obtained from $\{X_i\}_{i=1}^{N_1}$ by rotating on $\pi/2$ in the positive direction. This follows by comparing (8.37) and (8.38). The ε-biased estimators have the form:

$$\xi_1(x) = \left(\frac{4}{\pi}\right)^{N_1} (-1)^{\kappa_{N_1}} [\varphi_1(\bar{X}_{N_1}) \cos \psi_{N_1} + \varphi_2(\bar{X}_{N_1}) \sin \psi_{N_1}], \tag{8.44}$$

and

$$\xi_2(x) = \left(\frac{4}{\pi}\right)^{N_2} (-1)^{\kappa_{N_2}} [\varphi_1(\bar{y}_{N_2}) \cos \psi_{N_2} + \varphi_2(\bar{y}_{N_2}) \sin \psi_{N_2}]. \tag{8.45}$$

Note that the variance of the estimators is still divergent as $\varepsilon \to 0$. However, it is much less than that of the estimator (8.31).

8.5.4. Branching process

We consider now the general case of (9.9) with $p, q \geq 0$; $p + q = 1$.

Let us define a Markov branching process $\eta(x)$ starting at the point x, the center of the first circle $S(x, r_1)$. With probability p, the point y_1 is distributed on $S(x, r_1)$ uniformly, and with probability $q = 1 - p$, two points are sampled on $S(x, r_1)$, the first one being

$$y_1^+ = x + r_1 s^{(1)},$$

where $s^{(1)}$ has the distribution (8.39), and the second

$$y_1^- = x - r_1 s^{(1)}.$$

Next

$$y_2^+ = X_1 + r_2 \omega,$$

(ω isotropic) with probability p, and with probability $q = 1 - p$

$$y_2^\pm = X_1 \pm r_2 s^{(2)},$$

where X_1 is the previous state, i.e. it is y_1, y_1^+, or y_1^-, and $s^{(2)}$ has the distribution density (8.40) if the previous direction was not isotropic, and

(8.39) if it was isotropic, until the absorption in Γ_ε: y_k, $y_k^\pm \in \Gamma_\varepsilon$. Hence, after each isotropic scattering the transition (8.39) is used, continuing the scatterings according to (8.40), until an isotropic scattering happens again, etc.

Thus, one trajectory of $\eta(x)$ consists of one or two points in the first generation, 1, 2, 3 or 4 points of the second generation, etc. In the m-th generation we have $1, 2, \ldots,$ or 2^m points. Note that the set of the last (absorption) states lying in Γ_ε will consist of points of different generations.

Let us denote by N_{is} the number of the last states $\{y_k\}_{k=1}^{N_{is}} \in \Gamma_\varepsilon$ sampled isotropically, and by N_s the number of the last states $\{y_k^\pm\}_{k=1}^{N_s} \in \Gamma_\varepsilon$ sampled non-isotropically.

Let us choose now a fixed last state y_k or $y_k^\pm$, and let us consider the transitions leading from x to the last state. We denote by l_k the number of these transitions. Let $l_k^\pm$ be the number of transitions made according to the scheme

$$y_{j+1}^\pm = X_j \pm r_{j+1} s^{(j+1)}$$

inside these l_k steps.

We denote the trajectory of the branching process by

$$\eta_1 = \eta_1(x; \{y_k\}_{k=1}^{N_{is}}; \{y_k^\pm\}_{k=1}^{N_s})$$

indicating the starting point, the points $\{y_k\}_{k=1}^{N_{is}}$ absorbed in Γ_ε after an isotropic transition, and points $\{y_k^\pm\}_{k=1}^{N_s}$ absorbed in Γ_ε after a non-isotropic transition.

On η_1, we now define the following random estimator:

$$\hat{\zeta}_1(x) = \sum_{k=1}^{N_{is}} (-1)^{l_k^-} \left(\frac{2}{\pi}\right)^{l_k^+ + l_k^-} u_1(y_k) + \sum_{k=1}^{N_s} (-1)^{l_k^-} \left(\frac{2}{\pi}\right)^{l_k^+ + l_k^-} (\mathbf{u}(y_k^\pm) \cdot s^{(k+1)}),$$

where

$$(\mathbf{u}(y_k^\pm) \cdot s^{(k+1)}) = u_1(y_k^\pm) \cos \psi_k + u_2(y_k^\pm) \sin \psi_k.$$

By the construction,

$$u_1(x) = \mathbb{E}\hat{\zeta}_1(x)$$

for arbitrary $\varepsilon > 0$.

The ε-biased estimator has the form

$$\zeta_1(x) = \sum_{k=1}^{N_{is}} (-1)^{l_k^-} \left(\frac{2}{\pi}\right)^{l_k^+ + l_k^-} \varphi_1(\bar{y}_k) + \sum_{k=1}^{N_s} (-1)^{l_k^-} \left(\frac{2}{\pi}\right)^{l_k^+ + l_k^-} (\varphi(\bar{y}_k^{\pm}) \cdot s^{(k+1)}).$$

For the second component, we construct the trajectories $\eta_2(x)$ which are obtained from $\eta_1(x)$ by rotation through $\pi/2$ in the first step, repeating the rest of the trajectory. Let

$$\eta_2 = \eta_2(x; \{z_k\}_{k=1}^{N_{is}}; \{z_k^{\pm}\}_{k=1}^{N_s}),$$

and

$$\hat{\zeta}_2(x) = \sum_{k=1}^{N_{is}} (-1)^{l_k^-} \left(\frac{2}{\pi}\right)^{l_k^+ + l_k^-} u_2(z_k) + \sum_{k=1}^{N_s} (-1)^{l_k^-} \left(\frac{2}{\pi}\right)^{l_k^+ + l_k^-} (\mathbf{u}(z_k^{\pm}) \cdot s^{(k+1)}).$$

Then

$$u_2(x) = \mathbb{E}\hat{\zeta}_2(x)$$

for arbitrary $\varepsilon > 0$.

Remarkably, the estimators have the same form in the general n-dimensional case. For example, let us consider the 3-D case.

The branching process

$$\eta = \eta(x; \{y_k\}_{k=1}^{N_{is}}; \{y_k^{\pm}\}_{k=1}^{N_s})$$

is constructed according to the same scheme, using in the first sphere $\{s_3 > 0\}$ the density $p_+(s) = s_3/\pi$ and applying the corresponding rotation beginning from the second non-isotropic scattering.

The unbiased estimator in three dimensions has the form:

$$\hat{\zeta}_i(x) = \sum_{k=1}^{N_{is}} (-1)^{l_k^-} \left(\frac{3}{4}\right)^{l_k^- + l_k^+} u_i(y_k) + \sum_{k=1}^{N_s} (-1)^{l_k^-} \left(\frac{3}{4}\right)^{l_k^- + l_k^+} (\mathbf{u}(y_k^{\pm}) \cdot s^{(k+1)}),$$

and the corresponding ε-biased estimator is

$$\zeta_i(x) = \sum_{k=1}^{N_{is}} (-1)^{l_k^-} \left(\frac{3}{4}\right)^{l_k^- + l_k^+} \varphi_i(\bar{y}_k) + \sum_{k=1}^{N_s} (-1)^{l_k^-} \left(\frac{3}{4}\right)^{l_k^- + l_k^+} (\varphi(\bar{y}_k^{\pm}) \cdot s^{(k+1)}).$$

8.5.5. Analytical continuation with respect to the spectral parameter

There is another possibility for constructing the solution to the Lamé equation based on the analytical continuation of the solution to

$$\Delta^{*}\mathbf{u} - \nu\mathbf{u} = 0 \quad \mathbf{u}\Big|_{\Gamma} = \mathbf{g} \tag{8.46}$$

with respect to ν. Indeed, using the well known spectral properties of the Lamé equation, we can apply one of the numerical procedures of analytical continuation of the solution to (8.46) to the point $\nu = 0$. Thus it is necessary to construct a numerical procedure for the problem (8.46). We shall show that, for sufficiently large value of ν, the Random Walk on Spheres method converges. It means that for a fixed $\varepsilon > 0$ we can find ν depending on ε such that the Neumann series for the integral equation generated by the relevant spherical mean value relation for the problem (8.46) converges.

Let us rewrite the spherical mean value relation for (8.46) derived in Section 5.2 (see formula (5.33) in Theorem 5.7) as $\mathbf{u}(x) = K\mathbf{u}$, or

$$\mathbf{u}(x) = \frac{1}{4\pi} \int\limits_{S(x,R)} \frac{A\delta_{ij} + Bs_i s_j}{A\gamma_1(\eta) + B\gamma_2(\eta)} u_j(s)\, dS \ , \tag{8.47}$$

where

$$A = \gamma_1(\xi) - 2 + \gamma_2(\xi) - \gamma_2(\eta) \ ,$$
$$B = \gamma_1(\eta) - \gamma_1(\xi),$$

and

$$\gamma_1(\eta) = \frac{\operatorname{sh}\eta}{\eta}, \quad \gamma_2(\eta) = \frac{\operatorname{ch}\eta}{\eta^3} - \frac{\operatorname{sh}\eta}{\eta^3} \ .$$

Here $\eta = k_s R, \quad \xi = k_p R$.

We shall show now that for sufficiently large $\nu = \nu(\varepsilon)$, the norm $\|K\|_{L_\infty}$ of the integral operator generated by the spherical mean value relation (8.47) is less than 1. We first prove that

$$(a) \qquad \gamma_2(\eta) < 0, \quad A > 0 \ \text{ for } \ \eta > 0,$$

and

$$(b) \qquad B > 0 \ \text{ if } \ \xi > \frac{\ln g}{g-1} ,$$

where $g = \sqrt{2 + \lambda/\mu}$.

Indeed, since $\text{th}\,\eta < \eta$ for $\eta > 0$, we find

$$\gamma_2(\eta) = \frac{\text{sh}\,\eta - \eta\text{ch}\,\eta}{\eta^3} < 0 , \quad \eta > 0 .$$

Therefore, for $\eta > 0$, $\xi > 0$ we have

$$A = \gamma_1(\xi) - 2\gamma_2(\xi) - \gamma_2(\eta) > 0 .$$

Since $\eta = g\xi$, we can rewrite B as

$$B = \gamma_1(\eta) - \gamma_1(\xi) = \frac{\text{sh}\,g\xi}{g\xi} - \frac{\text{sh}\,\xi}{\xi} .$$

The inequality $B > 0$ implies $\text{sh}\,g\xi - g\text{sh}\,\xi > 0$, or

$$(e^{g\xi} - ge^{\xi}) + (ge^{-\xi} - e^{-g\xi}) > 0 ;$$

therefore,

$$e^{g\xi} - ge^{\xi} > 0 \quad \text{when } \xi > \frac{\ln g}{g-1} ,$$

which proves (b).

Let us estimate the norm

$$\|K\|_{L_\infty} = \max_{i,x} \sum_{j=1}^{3} |k_{ij}| ,$$

where

$$k_{ij} = \frac{A\delta_{ij} + Bs_i s_j}{A\gamma_1(\eta) + B\gamma_2(\eta)} .$$

We show that $\|K\|_{L_\infty} < 1$ if

$$\nu > \frac{(\lambda + 2\mu)\xi_0^2}{\varepsilon^2} , \tag{8.48}$$

where ξ_0 is the solution of the equation

$$\operatorname{sh}\xi - \frac{3\xi^2 + 2\xi e^{-\xi}}{\xi^2 - 2\xi + 2} = 0 .$$

Indeed,

$$\|K\|_{L_\infty} \leq \frac{|A| + 3|B|}{|A\gamma_1(\eta) + B\gamma_2(\eta)|} \leq \frac{|A| + 3|B|}{|A\gamma_1(\eta)| - |B\gamma_2(\eta)|} .$$

By the inequalities (a) and (b) we find

$$\|K\|_{L_\infty} \leq \frac{A + 3B}{A\gamma_1(\eta) + B\gamma_2(\eta)}$$

for $\xi > \ln g/(g-1)$, $\eta > 0$.

To ensure $\|K\|_{L_\infty} \leq 1$ we require

$$A + 3B < A\gamma_1(\eta) + B\gamma_2(\eta) ;$$

hence

$$A[\gamma_1(\eta) - 1] > B[3 - \gamma_2(\eta)] .$$

Substituting A and B into this equation yields

$$[\gamma_1(\xi) - 2\gamma_2(\xi) - \gamma_1(\eta)][\gamma_1(\eta) - 1] > [\gamma_1(\eta) - \gamma_1(\xi)][3 - \gamma_2(\eta)] .$$

Hence, for $\xi > 0$

$$\gamma_1(\eta) - 1 > \gamma_1(\eta) - \gamma_1(\xi) .$$

We find the values of ξ such that

$$\gamma_1(\xi) - 2\gamma_2(\xi) - \gamma_2(\eta) > 3 - \gamma_2(\eta),$$

or

$$\gamma_1(\xi) - 2\gamma_2(\xi) > 3 .$$

From this we get

$$\frac{\operatorname{sh}\xi}{\xi} - 2\left(\frac{\operatorname{ch}\xi}{\xi^2} - \frac{\operatorname{sh}\xi}{\xi^3}\right) > 3 .$$

Simple calculations lead to (8.48).

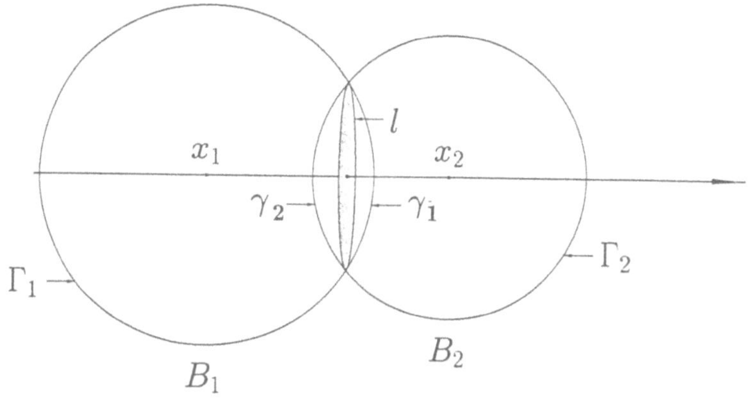

Figure 8.2. Two overlapping balls of the Schwarz iteration algorithm

8.6. Alternative Schwarz procedure

In this section we present a Random Walk method for solving the first boundary value problem for the Lamé equation

$$\begin{aligned} &\mu\Delta\mathbf{u} + (\lambda+\mu)\,\mathrm{grad}\,\mathrm{div}\,\mathbf{u} = 0, \quad x \in G, \\ &\mathbf{u}(y) = \mathbf{g}(y), \quad y \in \Gamma, \end{aligned} \tag{8.49}$$

where $G = B_1 \cup B_2$ is a union of two overlapping balls in $3D$

$$G = B_1(x_1, r_1) \cup B_2(x_2, r_2),$$

Γ is the boundary of G. Comparing with the general case of the Random Walk on Sphere process we see that in this special case $\varepsilon = 0$.

We introduce the following notation (see Fig. 8.2): γ_1 is the part of the sphere ∂B_1 lying in B_2, and γ_2 is the part of the sphere ∂B_2 lying in B_1, while $\Gamma_1 = \partial B_1 \setminus \gamma_1$, and $\Gamma_2 = \partial B_2 \setminus \gamma_2$. We denote the curve (a circle) of intersection of the spheres ∂B_1 and ∂B_2 by l. It is also convenient to introduce the notation: $\mathbf{u}|_{\Gamma_1} = \mathbf{g}^{(1)}$, and $\mathbf{u}|_{\Gamma_2} = \mathbf{g}^{(2)}$.

The generalized Poisson formula (see Theorem 6.3 in $2D$ and Theorem 6.7 in $3D$) for an arbitrary point $x \in B_1$ can be written in the matrix form as

$$\mathbf{u}(x) = (K_1 + F_1)\mathbf{u},$$

where

$$K_1\mathbf{u}(x) = \int_{\Gamma_1} C(x,y)\mathbf{g}(y)dS(y),$$

$$F_1\mathbf{u}(x) = \int_{\gamma_1} C(x,y)\mathbf{u}(y)dS(y),$$

and the matrix $C(x, y)$ is defined in the above mentioned theorems. The same relation is true at $y \in B_2$:

$$\mathbf{u}(y) = (K_2 + F_2)\mathbf{u},$$

where

$$K_2\mathbf{u}(y) = \int_{\Gamma_2} C(y,y')\mathbf{g}(y')dS(y'),$$

$$F_2\mathbf{u}(y) = \int_{\gamma_2} C(y,y')\mathbf{u}(y')dS(y').$$

Now, we can represent the solution through the iterative procedure

$$\mathbf{u}(x) = K_1\mathbf{g} + F_1K_2\mathbf{g} + F_1F_2K_1\mathbf{g} + F_1F_2F_1K_2\mathbf{g} + F_1F_2F_1F_2K_1\mathbf{g} + \dots .$$

which formally can be represented as

$$\mathbf{u}(x) = \sum_{j=1}^{\infty} G^j(K_l\mathbf{g}), \tag{8.50}$$

where

$$G^j = \underbrace{F_1F_2F_1F_2\dots F_1F_2\dots F_p}_{j \text{ times}},$$

$$l = \begin{cases} 1, & \text{if } j \text{ is even,} \\ 2 & \text{if } j \text{ is odd} \end{cases},$$

while

$$p = \begin{cases} 2, & \text{if } j \text{ is odd,} \\ 1 & \text{if } j \text{ is even.} \end{cases}$$

The multiple integrals in this sum can be evaluated by the Monte Carlo method on the corresponding Markov chain. For example, the term

$$F_1F_2F_1K_2\mathbf{g}(x) = \int_{\gamma_1} C(x,y_1)dS(y_1)\Big\{\int_{\gamma_2} C(y_1,y_2)dS(y_2) \\ \times \int_{\gamma_1} C(y_2,y_3)dS(y_3)\int_{\Gamma_2} C(y_3,y)\mathbf{g}(y)dS(y)\Big\}$$

is estimated on Markov chains starting at the point x with the transitions

$$x \to y_1 \in \gamma_1 \to y_2 \in \gamma_2 \to y_3 \in \gamma_3 \to y \in \Gamma_2 \ .$$

However we describe now a different iteration method, a generalized Schwarz method whose convergence was proved in [56].

Let us define the energy of deformation for any displacement vector $\mathbf{v}$ as

$$E(\mathbf{v}) = \int_G \Big[\lambda(\operatorname{div}\mathbf{v})^2 + \frac{\mu}{2}(\nabla^\alpha v_\beta + \nabla_\beta v^\alpha)^2\Big]dV \ . \tag{8.51}$$

It is well known that the boundary value problem (8.49) is equivalent to the problem of finding a function $\mathbf{v}$ such that $\mathbf{v}|_\Gamma = \mathbf{g}$ and which minimizes the energy integral (8.51).

The Schwarz iterative method consists in constructing a sequence of functions $\mathbf{u}^{(i)}$, $i = 0,1,\ldots$, such that the functions $\mathbf{u}^{(2k)}$ are regular in B_2 and $B_1 \setminus B_2$, satisfy in these domains the equation (8.49), they are continuous in G and

$$\mathbf{u}^{(2k)}|_{\Gamma_1} = \mathbf{g}^{(1)},\ \mathbf{u}^{(2k)}|_{\Gamma_2} = \mathbf{g}^{(2)},\ \mathbf{u}^{(2k)}|_{\gamma_2} = \begin{cases} \mathbf{g}^{(0)} & \text{if } k = 0\ , \\ \mathbf{u}^{(2k-1)} & \text{if } k > 0\ . \end{cases} \tag{8.52}$$

Here $\mathbf{g}^{(0)}$ is an arbitrary sufficiently smooth vector function whose values on the circle l coincide with those of the functions $\mathbf{g}^{(1)}$ and $\mathbf{g}^{(2)}$.

All functions $\mathbf{u}^{(2k+1)}$ are continuous in G, regular in B_1 and $B_2 \setminus B_1$, satisfy the equation (8.49), and

$$\mathbf{u}^{(2k+1)}|_{\Gamma_1} = \mathbf{g}^{(1)}, \ \mathbf{u}^{(2k+1)}|_{\Gamma_2} = \mathbf{g}^{(2)}, \ \mathbf{u}^{(2k+1)}|_{\gamma_1} = \mathbf{u}^{(2k)}|_{\gamma_1}, \quad k \geq 0. \quad (8.53)$$

The proof of convergence of $\mathbf{u}^{(k)}$ (see [56]) is based on the estimations

$$E(\mathbf{u}^{(2k)}) \leq E(\mathbf{u}^{(2k-1)})$$

and

$$E(\mathbf{u}^{(2k+1)}) \leq E(\mathbf{u}^{(2k)}) \ .$$

Finally, representing $\mathbf{u}^{(k)}$ through the Green function and taking the limit in the Green formula one finds that $\lim_{k\to\infty} \mathbf{u}^{(k)}$ solves the original problem. The same scheme works also if there are more than two overlapping balls.

Using the generalized Poisson formulae derived above we can construct an effective numerical algorithm based on the Schwarz method. Indeed, choose a sufficiently fine grid on the spheres ∂B_1 and ∂B_2, and calculate successively the functions $\mathbf{u}^{(k)}$ from the relevant Poisson formula. Note that in $3D$ some coefficients in the Poisson formula should be evaluated numerically in advance, for a fixed grid. In $2D$ all the coefficients are given explicitly.

Bibliography

[1] Aggarwala, B.D. (1974). On a Monte Carlo method for rectilinear plates. *Z. Angew. Math. und Mech.*, **54**, 194–196.

[2] Basheleyshvily, M.O. (1960). Analog of the Poisson formula in the elasticity theory. *Trudy Vych. Zentra AN GrSSR.*, **1**, 97–101.

[3] Budak, B.M., Samarsky, A.A. and Tykhonov, A.M. (1980). *Book of Exercises in Mathematical Physics.* Nauka, Moscow (in Russian).

[4] Courant, R. and Hilbert, D. (1989). *Methods of Mathematical Physics*, John Wiley, New York.

[5] Diaz, J. B. and Payne, L.E. (1958). On a mean value theorem, and its converse, for the displacement in the theory of elasticity. *Port. Mat.*, **17**, 123–126.

[6] Diaz, J. B. and Payne, L.E. (1958). Mean value theorems in the theory of elasticity. *Proc. Third Nat. Congr. Appl. Mech.*, 293–303.

[7] Dwight, H.B. (1961). *Tables of Integrals and other Mathematical Data.* Macmillan, New York.

[8] Elepov, B.S., Kronberg, A.A., Mikhailov, G.A. and Sabelfeld, K.K. (1980). *Solution of Boundary Value Problems by Monte Carlo Methods.* Nauka, Novosibirsk (in Russian).

[9] Ermakov, S.M., Nekrutkin, W.W. and Sipin, A.S. (1989). *Random Processes for Classical Equations of Mathematical Physics.* vol. **34** of Math. Appl. (Soviet Series). Kluwer, Dordrecht.

[10] Ermakov, S.M. and Mikhailov, G.A. (1982). *Statistical Simulation.* Nauka, Moscow (in Russian).

[11] Flatto, L. (1961). Functions with a mean value property. *J. Math. Mech.*, **10**, 216-238.

[12] Friedman, A. and Littman, W. (1962). Functions satisfying the mean value property. *Trans. Amer. Math.*, **102**.

[13] Fulks, W. (1964). An approximated Gauss mean value theorem. *Pacific J. Math.*, **14**, 513–516.

[14] Gopalsamy, K. and Aggarwala, B.D. (1970). Monte Carlo methods of some fourth order partial differential equations. *Z. Angew. Math. und Mech.*, **50**, 759–767.

[15] Gradstein, I.S. and Rygyk, I.M. (1971). *Tables of Integrals, Sums, Series and Products.* Nauka, Moscow (in Russian).

[16] Hersh, R. and Griegi, G (1971). Theory of random evolutions with applications to partial differential equations. *Trans. Am. Math. Soc.*, **156**, 405–418.

[17] John, F. (1955). *Plane Waves and Spherical Means.* Interscience, New York.

[18] Jörgens, K. (1970). *Linear Integral Operators.* Pitman, Boston.

[19] Korn, G. and Korn, T. (1977). *The Handbook on Mathematics.* Nauka, Moscow (in Russian).

[20] Krylov, N.V. (1967). On the Green function for the Dirichlet problem. *Uspehi Mat. Nauk* , **22**, No.2, 116-118 (in Russian).

[21] Kupradze, V.D., Gegelia, T.G., Basheleishvili, M.O. and Burkhuladze, T.V. (1976). *Three Dimensional Problems of Mathematical Elasticity Theory.* Nauka, Moscow (in Russian).

[22] Kuptsov, L.P. (1982). On expansions of some parabolic means. *Mat. Zametki*, **32**, No.4, 499–510 (in Russian).

[23] Leybenson, L.S. (1947). *Theory of Elasticity.* Gos. Izdad. Tekh-Theor. Lit., Moscow (in Russian).

[24] Markushevich, A.I. (1950). *Theory of Analytical Functions.* Gos. Izdad. Tekh-Theor. Lit., Moscow (in Russian).

[25] Mikhlin, S.G. (1952). *The problem of the minimum of the quadratic functional.* Gostekhizdat., Moscow.

[26] Miranda, C. (1970). *Partial Differential Equations of Elliptic Type.* Springer–Verlag, New York.

[27] Motoo, M. (1959). Some evaluations for continuous Monte Carlo method by using Brownian hitting process. *Ann. Stat. Math.*, **11**, 49–54.

[28] Muller, M.E. (1956). Some continuous Monte Carlo method for the Dirichlet problem. *Ann. Math. Stat.*, **27**, (3), 569-589.

[29] Naumov, V.V. (1987). Mean value relations for the equation of harmonic vibrations of an elastic body. In: Hydrodynamics of fast processes. Institute of Hydrodynamics. Novosibirsk, 147–153 (in Russian).

[30] Nicolescu, M. (1936). *Les fonctions polyharmoniques. Act. Sc.*, 331, Hermann.

[31] Pagani, C. (1978). On elliptic–parabolic equations of second order and approximating integral equations. *Complex Analysis and Applications.*, 471–493. Nauka, Moscow.

[32] Pizetti, P. (1909). Sulla media deivalori che una funzione dei punti dello spazio assume alla superficie di una sfera. *Rend. Lincei*, **5**, No.18, 309–316.

[33] Pologiy, G.N. (1964). *Equations of Mathematical Physics.* Vys. Shkola, Moscow (in Russian).

[34] Poritsky, H. (1938). Generalizations of the Gauss law of the spherical mean. *Trans.Amer. Math. Soc.*, **43**, 199–225.

[35] Pucci, C. and Talenti, G. (1976). Elliptic (second-order) partial differential equations with measurable coefficients and approximating integral equations. *Advances in Mathematics.*, **19**, 48–105.

[36] Roux, R. (1987). Generalized Brownian motion and elasticity. *J. Stat. Physics*, **48**, 201–213.

[37] Sabelfeld, K.K. (1982). Vector Monte Carlo algorithms for solving systems of elliptic equations of the second order and the Lamé equation. *Dokl.Akad.Nauk SSSR*, **262**, No.5, 1076–1080 (in Russian).

[38] Sabelfeld, K.K. (1983). The Balayage and averaging principles for constructing Monte Carlo methods for boundary value problems. *Journ. of Comp. Mathem. and Mathem. Physics.*, **23**, 366–379 (in Russian).

[39] Sabelfeld, K.K. (1987). Random Walk inside the domain for solving boundary value problems. *Soviet J. on Numerical analysis and mathem. modelling.* **2**, No.3, 209-237.

[40] Sabelfeld, K.K. (1991). *Monte Carlo Methods in Boundary Value Problems.* Springer-Verlag, Berlin.

[41] Sabelfeld, K.K. (1996). Integral and probabilistic representations for systems of elliptic equations. *Mathem. and Computer Modelling*, **23**, No.8/9, 111–129.

[42] Sabelfeld, K.K. and Chentsov, N.N. (1986). Vector and scalar random walk algorithms for solving boundary value problems. First Internat. Sympos. of the Bernoulli society, Tashkent, 1986, 461.

[43] Sabelfeld, K.K. and Shalimova, I.A. (1987). The mean value theorem for elliptic equations and thermoelastic wave equation. In: *Numerical methods of statistical simulation* , Novosibirsk, Computing Center, 88-94 (in Russian)

[44] Sabelfeld, K.K. and Shalimova, I.A. (1988). Mean value theorems in Monte Carlo methods. *Sov. J. Num. Anal. Math. Model.*, **3**, No.3., 217-230.

[45] Sabelfeld, K.K. and Shalimova, I.A. (1995). Random Walk on Spheres Process for exterior Dirichlet problem. *Monte Carlo Methods and Applications.*, **1**, No.4, 325–331.

[46] Sabelfeld, K.K. and Simonov, N.A. (1994). *Random Walks on Boundary for Solving PDEs.* VSP, Utrecht, The Netherlands.

[47] Sabelfeld, K.K. and Talay, D. (1995). Integral formulation of the boundary value problems and the method of random walk on spheres. *Monte Carlo Methods and Applications*, **1**, 1–34.

[48] Sato, T. (1956). Sur le problème de Dirichlet généralizé pour l'équation $\Delta u = f(P, u, \partial u)$. *Composito Mat.*, **14**, 112–167.

[49] Sbrana, F. (1952). Una proprieta caratteristica delle equazioni dell' elasticita. *Quaderno scientifico del Convegno matematico di Modena.*, 84–88.

[50] Schulze, B.W. and Wildenhain, G. (1977). *Methoden der Potentialtheorie für elliptische Differentialgleichungen beliebiger Ordnung.* Akademie-Verlag, Berlin.

[51] Shalimova, I.A. (1989). A mean-value theorem at a noncentral point for the Lamé system. In: *Theory and Applications of Statistical Modeling.* Computing Centre, Novosibirsk, p.135–146 (in Russian).

[52] Shalimova, I.A. (1993). Spherical mean-value theorem at a noncentral point for the Lamé system and its use for constructing Monte Carlo algorithms. *Rus. J. Num. Anal. Math. Model.*, **8**, No.4, 327–340.

[53] Shalimova, I.A. (1993). Mean value theorems in the elasticity theory. In: *Proced. Comp. Centre, ser. Numerical Mathematics.* Computing Centre, Novosibirsk, 85–106 (in Russian).

[54] Sobol, I.M. (1973). *Numerical Monte Carlo Methods.* Nauka, Moscow (in Russian).

[55] Sobolev, S.L. (1974). *Introduction to Cubic Formulas.* Nauka, Moscow (in Russian).

[56] Sobolev, S.L. (1936). The Schwarz algorithm in elasticity theory. *Doklady AN SSSR*, **4**, 235–238 (in Russian).

[57] Sobolev, S.L. (1937). On a boundary value problem for polyharmonic functions. *Matem. sbornik*, **2**(44), 467–500 (in Russian).

[58] Wasow, W.R. and Forsythe, G.E. (1960). *Finite-Difference Methods for Partial Differential Equations.* Wiley, New York.

[59] Zalcman, L. (1973). Mean values and differential equations. *Israel Journal of Mathematics*, **14**, 4, 339–352.

Bei Fragen zur Produktsicherheit wenden Sie sich bitte an:
If you have any questions regarding product safety,
please contact:

Walter de Gruyter GmbH
Genthiner Straße 13
10785 Berlin
productsafety@degruyterbrill.com